2018

INTERNATIONAL CODES®

IPSDC®

A Member of the International Code Family®

INTERNATIONAL PRIVATE SEWAGE DISPOSAL CODE®

NOWRA
National Onsite Wastewater Recycling Association

ICC
INTERNATIONAL CODE COUNCIL®

2018 International Private Sewage Disposal Code®

First Printing: August 2017

ISBN: 978-1-60983-747-1 (soft-cover edition)

Date of First Publication: August 31, 2017

PRINTED IN THE U.S.A.

PREFACE

Introduction

The *International Private Sewage Disposal Code*® (IPSDC®), establishes minimum requirements for sewage disposal systems using prescriptive and performance-related provisions. It is founded on broad-based principles that make possible the use of new materials and new sewage disposal designs. This 2018 edition is fully compatible with all of the *International Codes*® (I-Codes®) published by the International Code Council® (ICC®), including the *International Building Code*®, *International Energy Conservation Code*®, *International Existing Building Code*®, *International Fire Code*®, *International Fuel Gas Code*®, *International Green Construction Code*®, *International Mechanical Code*®, *International Plumbing Code*®, *International Property Maintenance Code*®, *International Residential Code*®, *International Swimming Pool and Spa Code*®, *International Wildland-Urban Interface Code*®, *International Zoning Code*® and *International Code Council Performance Code*®.

The I-Codes, including this *International Private Sewage Disposal Code*, are used in a variety of ways in both the public and private sectors. Most industry professionals are familiar with the I-Codes as the basis of laws and regulations in communities across the U.S. and in other countries. However, the impact of the codes extends well beyond the regulatory arena, as they are used in a variety of nonregulatory settings, including:

- Voluntary compliance programs such as those promoting sustainability, energy efficiency and disaster resistance.
- The insurance industry, to estimate and manage risk, and as a tool in underwriting and rate decisions.
- Certification and credentialing of individuals involved in the fields of building design, construction and safety.
- Certification of building and construction-related products.
- U.S. federal agencies, to guide construction in an array of government-owned properties.
- Facilities management.
- "Best practices" benchmarks for designers and builders, including those who are engaged in projects in jurisdictions that do not have a formal regulatory system or a governmental enforcement mechanism.
- College, university and professional school textbooks and curricula.
- Reference works related to building design and construction.

In addition to the codes themselves, the code development process brings together building professionals on a regular basis. It provides an international forum for discussion and deliberation about building design, construction methods, safety, performance requirements, technological advances and innovative products.

Development

This 2018 edition presents the code as originally issued, with changes reflected in the 1997 through 2015 editions and further changes approved by the ICC Code Development Process through 2017. A new edition such as this is promulgated every 3 years.

This code is founded on principles intended to establish provisions consistent with the scope of a sewage disposal code that adequately protects public health, safety and welfare; provisions that do not unnecessarily increase construction costs; provisions that do not restrict the use of new materials, products or methods of construction and provisions that do not give preferential treatment to particular types or classes of materials, products or methods of construction.

Maintenance

The *International Private Sewage Disposal Code* is kept up to date through the review of proposed changes submitted by code enforcement officials, industry representatives, design professionals and other interested parties. Proposed changes are carefully considered through an open code development process in which all interested and affected parties may participate.

The ICC Code Development Process reflects principles of openness, transparency, balance, due process and consensus, the principles embodied in OMB Circular A-119, which governs the federal government's use of private-sector standards. The ICC process is open to anyone; there is no cost to participate, and people can participate without travel cost through the ICC's cloud-based app, cdpAccess®. A broad cross section of interests are represented in the ICC Code Development Process. The codes, which are updated regularly, include safeguards that allow for emergency action when required for health and safety reasons.

In order to ensure that organizations with a direct and material interest in the codes have a voice in the process, the ICC has developed partnerships with key industry segments that support the ICC's important public safety mission. Some code development committee members were nominated by the following industry partners and approved by the ICC Board:

- American Society of Plumbing Engineers (ASPE)
- National Association of Home Builders (NAHB)
- Plumbing Heating and Cooling Contractors (PHCC)
- American Institute of Architects (AIA)

The code development committees evaluate and make recommendations regarding proposed changes to the codes. Their recommendations are then subject to public comment and council-wide votes. The ICC's governmental members—public safety officials who have no financial or business interest in the outcome—cast the final votes on proposed changes.

The contents of this work are subject to change through the code development cycles and by any governmental entity that enacts the code into law. For more information regarding the code development process, contact the Codes and Standards Development Department of the International Code Council.

While the I-Code development procedure is thorough and comprehensive, the ICC, its members and those participating in the development of the codes disclaim any liability resulting from the publication or use of the I-Codes, or from compliance or noncompliance with their provisions. The ICC does not have the power or authority to police or enforce compliance with the contents of this code.

Code Development Committee Responsibilities (Letter Designations in Front of Section Numbers)

In each code development cycle, code change proposals to this code are considered at the Committee Action Hearings by the International Plumbing Code Development Committee, whose action constitutes a recommendation to the voting membership for final action on the proposed change. Code change proposals to sections of the code that are preceded by a bracketed letter designation, such as [A], will be considered by a committee other than the Plumbing Code Development Committee. For example, proposed changes to Section [BS] 303.1 will be considered by the IBC—Structural Code Development Committee at the Committee Action Hearings in the 2019 (Group B) code development cycle.

The bracketed letter designations for committees responsible for portions of this code are as follows:

[A] = Administrative Code Development Committee;

[BS] = IBC—Structural Code Development Committee; and

[P] = International Plumbing Code Development Committee.

For the development of the 2021 edition of the I-Codes, there will be two groups of code development committees and they will meet in separate years. Note that these are tentative groupings.

Group A Codes (Heard in 2018, Code Change Proposals Deadline: January 8, 2018)	Group B Codes (Heard in 2019, Code Change Proposals Deadline: January 7, 2019)
International Building Code – Egress (Chapters 10, 11, Appendix E) – Fire Safety (Chapters 7, 8, 9, 14, 26) – General (Chapters 2–6, 12, 27–33, Appendices A, B, C, D, K, N)	Administrative Provisions (Chapter 1 of all codes except IECC, IRC and IgCC, administrative updates to currently referenced standards, and designated definitions)
International Fire Code	**International Building Code** – Structural (Chapters 15–25, Appendices F, G, H, I, J, L, M)
International Fuel Gas Code	**International Existing Building Code**
International Mechanical Code	**International Energy Conservation Code—Commercial**
International Plumbing Code	**International Energy Conservation Code—Residential** – IECC—Residential – IRC—Energy (Chapter 11)
International Property Maintenance Code	**International Green Construction Code** (Chapter 1)
International Private Sewage Disposal Code	**International Residential Code** – IRC—Building (Chapters 1–10, Appendices E, F, H, J, K, L, M, O, Q, R, S, T)
International Residential Code – IRC—Mechanical (Chapters 12–23) – IRC—Plumbing (Chapters 25–33, Appendices G, I, N, P)	
International Swimming Pool and Spa Code	
International Wildland-Urban Interface Code	
International Zoning Code	
Note: Proposed changes to the ICC *Performance Code*™ will be heard by the code development committee noted in brackets [ ] in the text of the ICC *Performance Code*™.	

Code change proposals submitted for code sections that have a letter designation in front of them will be heard by the respective committee responsible for such code sections. Because different committees hold Committee Action Hearings in different years, proposals for this code will be heard by committees in both the 2018 (Group A) and 2019 (Group B) code development cycles.

Note that every section of Chapter 1 of this code is designated as the responsibility of the Administrative Code Development Committee, and that committee is part of the Group B portion of the code hearings. This committee will hold its code development hearings in 2019 to consider all code change proposals for Chapter 1 of this code and proposals for Chapter 1 of all I-Codes except the *International Energy Conservation Code, International Residential Code* and *International Green Construction Code*. Therefore, any proposals received for Chapter 1 of this code will be assigned to the Administrative Code Development Committee for consideration in 2019.

It is very important that anyone submitting code change proposals understand which code development committee is responsible for the section of the code that is the subject of the code change proposal. For further information on the code development committee responsibilities, please visit the ICC website at www.iccsafe.org/scoping.

Marginal Markings

Solid vertical lines in the margins within the body of the code indicate a technical change from the requirements of the 2015 edition. Deletion indicators in the form of an arrow (➡) are provided in the margin where an entire section, paragraph, exception or table has been deleted or an item in a list of items or a table has been deleted.

Coordination of the International Codes

The coordination of technical provisions is one of the strengths of the ICC family of model codes. The codes can be used as a complete set of complementary documents, which will provide users with full integration and coordination of technical provisions. Individual codes can also be used in subsets or as stand-alone documents. To make sure that each individual code is as complete as possible, some technical provisions that are relevant to more than one subject area are duplicated in some of the model codes. This allows users maximum flexibility in their application of the I-Codes.

Italicized Terms

Words and terms defined in Chapter 2, Definitions, are *italicized* where they appear in code text and the Chapter 2 definition applies. Where such words and terms are not italicized, common-use definitions apply. The words and terms selected have code-specific definitions that the user should read carefully to facilitate better understanding of the code.

Adoption

The International Code Council maintains a copyright in all of its codes and standards. Maintaining copyright allows ICC to fund its mission through sales of books, in both print and electronic formats. The ICC welcomes adoption of its codes by jurisdictions that recognize and acknowledge the ICC's copyright in the code, and further acknowledge the substantial shared value of the public/private partnership for code development between jurisdictions and the ICC.

The ICC also recognizes the need for jurisdictions to make laws available to the public. All I-Codes and I-Standards, along with the laws of many jurisdictions, are available for free in a nondownloadable form on the ICC's website. Jurisdictions should contact the ICC at adoptions@iccsafe.org to learn how to adopt and distribute laws based on the *International Private Sewage Disposal Code* in a manner that provides necessary access, while maintaining the ICC's copyright.

Section 101.1. Insert: [NAME OF JURISDICTION]

Section 106.4.2. Insert: [APPROPRIATE SCHEDULE]

Section 106.4.3. Insert: [PERCENTAGES IN TWO LOCATIONS]

Section 108.4. Insert: [OFFENSE, DOLLAR AMOUNT, NUMBER OF DAYS]

Section 108.5. Insert: [DOLLAR AMOUNT IN TWO LOCATIONS]

Section 405.2.5. Insert: [DATE IN THREE LOCATIONS]

Section 405.2.6. Insert: [DATE IN TWO LOCATIONS]

EFFECTIVE USE OF THE INTERNATIONAL PRIVATE SEWAGE DISPOSAL CODE

The *International Private Sewage Disposal Code* (IPSDC) is a model code that regulates minimum requirements for the installation of new or the alteration of existing private sewage disposal systems. Where a building cannot be served by a public sewer system, the building site must be provided with a system for treating the wastewater generated from the use of plumbing fixtures in the building. The IPSDC addresses site evaluations, materials, various soil absorption systems, holding tanks, cesspools and on-site wastewater treatment systems. The IPSDC provides a total approach for the on-site, safe disposal of the waste flow discharged to the plumbing fixtures in a building.

The IPSDC is a specification- (prescriptive-) oriented code with very few occurrences of performance-oriented text. The site soil must be evaluated in a prescribed manner to determine its ability to accept the waste flow. The chosen waste treatment method must be designed in a prescribed manner for the soil conditions at the building site, constructed using prescribed materials and installed according to prescribed dimensions. The IPSDC sets forth the minimum acceptable requirements for private sewage disposal systems in order to protect humans and the environment from insanitary conditions that would develop if waste flows were not rendered harmless.

Arrangement and Format of the 2018 IPSDC

The format of the IPSDC allows each chapter to be devoted to a particular subject with the exception of Chapter 3, which contains general subject matters that are not extensive enough to warrant their own independent chapter. The IPSDC is divided into 11 different parts:

Chapters	Subjects
1-2	Administration and Definitions
3	General Regulations
4	Site Evaluation and Requirements
5	Materials
6, 7, 9 & 10	Effluent Absorption and Distribution Systems
8	Tanks
11	Wastewater Treatment Systems
12	Inspections
13	Nonliquid Saturated Treatment Systems
14	Referenced Standards
Appendices A & B	Appendices

The following is a chapter-by-chapter synopsis of the scope and intent of the provisions of the *International Private Sewage Disposal Code*:

Chapter 1 Scope and Administration. This chapter contains provisions for the application, enforcement and administration of subsequent requirements of the code. In addition to establishing the scope of the code, Chapter 1 identifies which buildings and structures come under its purview. Chapter 1 is largely concerned with maintaining "due process of law" in enforcing the requirements contained in the body of this code. Only through careful observation of the administrative provisions can the code official reasonably expect to demonstrate that "equal protection under the law" has been provided.

Chapter 2 Definitions. Chapter 2 is the repository of the definitions of terms used in the body of the code. Codes are technical documents and every word, term and punctuation mark can impact the meaning of the code text and the intended results. The code often uses terms that have a unique meaning in the code and the code meaning can differ substantially from the ordinarily understood meaning of the term as used outside of the code.

The terms defined in Chapter 2 are deemed to be of prime importance in establishing the meaning and intent of the code text. The user of the code should be familiar with and consult this chapter because the definitions are essential to the correct interpretation of the code and the user may not be aware that a term is defined.

Where understanding of a term's definition is especially key to or necessary for understanding of a particular code provision, the term is shown in *italics*. This is true only for those terms that have a meaning that is unique to the code. In other words, the generally understood meaning of a term or phrase might not be sufficient or consistent with the meaning prescribed by the code; therefore, it is essential that the code-defined meaning be known.

Guidance regarding tense, gender and plurality of defined terms, as well as guidance regarding terms not defined in this code, is provided.

Chapter 3 General Regulations. The content of Chapter 3 is often referred to as "miscellaneous," rather than general regulations. Chapter 3 received that label because it is the only chapter in the code where requirements do not interrelate. If a requirement cannot be located in another chapter, it can be found in this chapter. Specific requirements concerning flood hazard areas are in this chapter.

Chapter 4 Site Evaluation and Requirements. A private sewage disposal system has an effluent that cannot be directly discharged into waterways or open ponds. Soil of the right consistency and water content provides natural filtering and treatment of this discharge. Because soil conditions vary widely, even on the same building site, tests and inspections of the soils must be performed to evaluate the degree to which the soil can accept these liquids. The results of the tests provide necessary information to design an adequate private sewage disposal system. Chapter 4 provides the methods for evaluating the building site.

Chapter 5 Materials. Private sewage disposal systems depend on the strength, quality and chemical resistance of the components that make up the system. To that end, the purpose of Chapter 5 is to specify the minimum material and component standards to ensure that the private sewage disposal system will correctly perform for its intended life.

Chapter 6 Soil Absorption Systems. The design of soil absorption systems depends heavily on the result of the tests and evaluation of the site soil conditions required in Chapter 4. Where soil is less permeable, the area of the soil absorption must be large as compared to that required for soils that are highly permeable. The type of building that is being served by the private sewage disposal system also affects the size of the planned soil absorption area. Chapter 6 provides the methods for computing the required absorption area and details for the proper installation of the soil absorption systems.

Chapter 7 Pressure Distribution Systems. Chapter 6 deals with gravity-type soil absorption systems or systems where the effluent is allowed to drain out of the distribution piping by gravity. Chapter 7 offers an alternative method of discharging the effluent into the ground by pressure means. As such, Chapter 7 provides the necessary details for designing the piping and pumping systems for pressure distribution systems.

Chapter 8 Tanks. Tanks are an integral part of any private sewage disposal system whether they serve as treatment (septic) tanks or merely just holding tanks for leveling the peaks in flow to the system. Where tanks are used for treatment, the dimensions, volume and location of internal features are very important to ensure that the solid wastes are kept within the tank so as to not clog the effluent distribution system. Where tanks are used for holding purposes, they must be sized large enough to accommodate the total of peak flows coming from a building. Chapter 8 provides the necessary requirements for tanks.

Chapter 9 Mound Systems. Mound systems are another method for applying the effluent from a private sewage disposal system to the soil. This type of system may be advantageous in some localities due to the existing soil conditions. Chapter 9 has specific requirements for soil and site evaluations for mound systems.

Chapter 10 Cesspools. Although prohibited from being installed as a permanent private sewage disposal system, cesspools may be necessary where permanent systems are under repair, or are being built. Chapter 10 provides the details for constructing a cesspool.

Chapter 11 Residential Wastewater Systems. Another method of private sewage disposal is a small wastewater treatment plant. Where permitted, these systems can discharge effluent directly to streams and rivers. Chapter 11 specifies the standard to which wastewater treatment plants must conform.

Chapter 12 Inspections. The best soil and site analysis along with the best design will be rendered useless if the system is not installed according to the plans for the system. Chapter 12 provides requirements for inspection of private sewage disposal systems.

Chapter 13 Nonliquid Saturated Treatment Systems. In some locations, water for the flushing of waste into and through a sanitary piping system is not available. For example, a toilet facility provided for a remote campground without running water would require such a system. Chapter 13 specifies the standard to which nonliquid saturated treatment systems must conform.

Chapter 14 Referenced Standards. Chapter 14 contains a comprehensive list of all standards that are referenced in the code. The standards are part of the code to the extent of the reference to the standard. Compliance with the referenced standard is necessary for compliance with this code. By providing specifically adopted standards, the construction and installation requirements necessary for compliance with the code can be readily determined. The basis for code compliance is, therefore, established and available on an equal basis to the code official, contractor, designer and owner.

Chapter 14 is organized in a manner that makes it easy to locate specific standards. It lists all of the referenced standards, alphabetically, by acronym of the promulgating agency of the standard. Each agency's standards are then listed in either alphabetical or numeric order based on the standard identification. The list also contains the title of the standard; the edition (date) of the standard referenced; any addenda included as part of the ICC adoption; and the section or sections of this code that reference the standard.

Appendix A System Layout Illustrations. Because each chapter of this code uses only words to describe requirements, illustrations can offer greater insight as to what the words mean. Appendix A has a number of illustrations referenced to specific sections of the code to help the reader gain a better understanding of the code's requirements.

Appendix B Tables for Pressure Distribution Systems. The design of a pressure distribution system is accomplished by the use of several complex formulas found in Chapter 7. Because a user of the code may not have the necessary experience to manipulate the formulas, a tabular approach for designing pressure distribution systems is provided in Appendix B.

TABLE OF CONTENTS

CHAPTER 1

SCOPE AND ADMINISTRATION

User note:

About this chapter: *Chapter 1 establishes the limits of applicability of this code and describes how the code is to be applied and enforced. Chapter 1 is in two parts: Part 1—Scope and Application (Sections 101–102) and Part 2—Administration and Enforcement (Sections 103–110). Section 101 identifies which buildings and structures come under its purview and references other I-Codes as applicable. Standards and codes are scoped to the extent referenced (see Section 102.10).*

This code is intended to be adopted as a legally enforceable document and it cannot be effective without adequate provisions for its administration and enforcement. The provisions of Chapter 1 establish the authority and duties of the code official appointed by the authority having jurisdiction and also establish the rights and privileges of the design professional, contractor and property owner.

PART 1—SCOPE AND APPLICATION

SECTION 101
GENERAL

[A] 101.1 Title. These regulations shall be known as the *Private Sewage Disposal Code* of **[NAME OF JURISDICTION]** hereinafter referred to as "this code."

[A] 101.2 Scope. Septic tank and effluent absorption systems or other treatment tank and effluent disposal systems shall be permitted where a public sewer is not available to the property served. Unless specifically approved, the *private sewage disposal system* of each building shall be entirely separate from and independent of any other building. The use of a common system or a system on a parcel other than the parcel where the structure is located shall be subject to the full requirements of this code as for systems serving public buildings.

[A] 101.2.1 Appendices. Provisions in the appendices shall not apply unless specifically adopted

[A] 101.3 Public sewer connection. Where public sewers become available to the premises served, the use of the *private sewage disposal system* shall be discontinued within that period of time required by law, but such period shall not exceed one year. The building sewer shall be disconnected from the *private sewage disposal system* and connected to the public sewer.

[A] 101.4 Abandoned systems. Abandoned *private sewage disposal systems* shall be plugged or capped in an approved manner. Abandoned treatment tanks and *seepage pits* shall have the contents pumped and discarded in an approved manner. The top or entire tank shall be removed and the remaining portion of the tank or excavation shall be filled immediately.

[A] 101.5 Failing system. When a *private sewage disposal system* fails or malfunctions, the system shall be corrected or use of the system shall be discontinued within that period of time required by the code official, but such period shall not exceed one year.

[A] 101.5.1 Failure. A failing *private sewage disposal system* shall be one causing or resulting in any of the following conditions:

1. The failure to accept sewage discharge and backup of sewage into the structure served by the *private sewage disposal system.*
2. The discharge of sewage to the surface of the ground or to a drain tile.
3. The discharge of sewage to any surface or ground water.
4. The introduction of sewage into saturation zones adversely affecting the operation of a *private sewage disposal system.*

[A] 101.6 Intent. The purpose of this code is to establish minimum standards to provide a reasonable level of safety health, property protection and public welfare by regulating and controlling the design, construction, installation, quality of materials, location, operation and maintenance or use of *private sewage disposal systems.*

[A] 101.7 Severability. If any section, subsection, sentence, clause or phrase of this code is for any reason held to be unconstitutional, such decision shall not affect the validity of the remaining portions of this code.

SECTION 102
APPLICABILITY

[A] 102.1 General. Where there is a conflict between a general requirement and a specific requirement, the specific requirement shall govern. Where, in any specific case, different sections of this code specify different materials, methods of construction or other requirements, the most restrictive shall govern.

[A] 102.2 Other laws. The provisions of this code shall not be deemed to nullify any provisions of local, state or federal law.

[A] 102.3 Application of references. Reference to chapter section numbers, or to provisions not specifically identified by number, shall be construed to refer to such chapter, section or provision of this code.

[A] 102.4 Existing installations. *Private sewage disposal systems* lawfully in existence at the time of the adoption of this code shall be permitted to have their use and maintenance continued if the use, maintenance or repair is in accordance with the original design and no hazard to life, health or property is created by the system.

[A] 102.5 Maintenance. *Private sewage disposal systems*, materials and appurtenances, both existing and new, and all parts thereof shall be maintained in proper operating condition in accordance with the original design in a safe and sanitary condition. Devices or safeguards that are required by this code shall be maintained in compliance with the edition of the code under which they were installed. The owner or the owner's authorized agent shall be responsible for maintenance of *private sewage disposal systems*. To determine compliance with this provision, the code official shall have the authority to require reinspection of any *private sewage disposal system.*

[A] 102.6 Additions, alterations or repairs. Additions, alterations, renovations or repairs to any *private sewage disposal system* shall conform to that required for a new system without requiring the existing system to comply with all the requirements of this code. Additions, alterations or repairs shall not cause an existing system to become unsafe, insanitary or overloaded.

Minor additions, alterations, renovations and repairs to existing systems shall meet the provisions for new construction, unless such work is done in the same manner and arrangement as was in the existing system, is not hazardous and is approved.

[A] 102.7 Change in occupancy. It shall be unlawful to make any change in the occupancy of any structure that will subject the structure to any special provision of this code applicable to the new occupancy without approval of the code official. The code official shall certify that such structure meets the intent of the provisions of law governing building construction for the proposed new occupancy and that such change of occupancy does not result in any hazard to the public health, safety or welfare.

[A] 102.8 Historic buildings. The provisions of this code relating to the construction, alteration, repair, enlargement, restoration, relocation or moving of buildings or structures shall not be mandatory for existing buildings or structures identified and classified by the state or local jurisdiction as historic buildings when such buildings or structures are judged by the code official to be safe and in the public interest of health, safety and welfare regarding any proposed construction, alteration, repair, enlargement, restoration, relocation or moving of buildings.

[A] 102.9 Moved buildings. Except as determined by Section 102.4, *private sewage disposal systems* that are a part of buildings or structures moved into or within the jurisdiction shall comply with the provisions of this code for new installations.

[A] 102.10 Referenced codes and standards. The codes and standards referenced in this code shall be those that are listed in Chapter 14 and such codes and standards shall be considered to be part of the requirements of this code to the prescribed extent of each such reference and as further regulated in Sections 102.10.1 and 102.10.2.

Exception: Where enforcement of a code provision would violate the conditions of the listing of the equipment or appliance, the conditions of the listing and the manufacturer's installation instructions shall apply.

[A] 102.10.1 Conflicts. Where conflicts occur between provisions of this code and the referenced standards, the provisions of this code shall apply.

[A] 102.10.2 Provisions in referenced codes and standards. Where the extent of the reference to a referenced code or standard includes subject matter that is within the scope of this code, the provisions of this code, as applicable, shall take precedence over the provisions in the referenced code or standard.

[A] 102.11 Requirements not covered by code. Any requirements necessary for the proper operation of an existing or proposed *private sewage disposal system*, or for the public safety, health and general welfare, not specifically covered by this code, shall be determined by the code official.

PART 2—ADMINISTRATION AND ENFORCEMENT

SECTION 103
DEPARTMENT OF
PRIVATE SEWAGE DISPOSAL INSPECTION

[A] 103.1 General. The Department of Private Sewage Disposal Inspection is hereby created and the executive official in charge thereof shall be known as the code official.

[A] 103.2 Appointment. The code official shall be appointed by the chief appointing authority of the jurisdiction.

[A] 103.3 Deputies. In accordance with the prescribed procedures of the jurisdiction and with the concurrence of the appointing authority, the code official shall have the authority to appoint a deputy code official, other related technical officers, inspectors and other employees. Such employees shall have powers as delegated by the code official.

[A] 103.4 Liability. The code official, member of the board of appeals or employee charged with the enforcement of this code, while acting for the jurisdiction in good faith and without malice in the discharge of the duties required by this code or other pertinent law or ordinance, shall not thereby be rendered civilly or criminally liable personally, and is hereby relieved from all personal liability for any damage accruing to persons or property as a result of any act or by reason of an act or omission in the discharge of official duties.

[A] 103.4.1 Legal defense. Any suit or criminal complaint instituted against any officer or employee because of an act performed by that officer or employee in the lawful discharge of duties and under the provisions of this code shall be defended by the legal representatives of the jurisdiction until the final termination of the proceedings. The code official or any subordinate shall not be liable for costs in any action, suit or proceeding that is instituted in pursuance of the provisions of this code.

SECTION 104
DUTIES AND POWERS OF THE CODE OFFICIAL

[A] 104.1 General. The code official is hereby authorized and directed to enforce the provisions of this code. The code official shall have the authority to render interpretations of this code and to adopt policies and procedures in order to clarify the application of its provisions. Such interpretations, policies and procedures shall be in compliance with the intent and purpose of this code. Such policies and procedures shall not have the effect of waiving requirements specifically provided for in this code.

[A] 104.2 Applications and permits. The code official shall receive applications, review construction documents and issue permits for the installation and alteration of *private sewage disposal systems*, inspect the premises for which such permits have been issued and enforce compliance with the provisions of this code.

[A] 104.3 Inspections. The code official shall make all of the required inspections, or shall accept reports of inspection by approved agencies or individuals. Reports of such inspections shall be in writing and be certified by a responsible officer of such approved agency or by the responsible individual. The code official is authorized to engage such expert opinion as deemed necessary to report on unusual technical issues that arise, subject to the approval of the appointing authority.

[A] 104.4 Right of entry. Where it is necessary to make an inspection to enforce the provisions of this code, or where the code official has reasonable cause to believe that there exists in any building or on any premises any conditions or violations of this code that make the building or premises unsafe, insanitary, dangerous or hazardous, the code official shall have the authority to enter the building or premises at all reasonable times to inspect or to perform the duties imposed on the code official by this code. If such building or premises is occupied, the code official shall present credentials to the occupant and request entry. If such building or premises is unoccupied, the code official shall first make a reasonable effort to locate the owner, the owner's authorized agent or other person having charge or control of the building or premises and request entry. If entry is refused, the code official has recourse to every remedy provided by law to secure entry.

Where the code official shall have first obtained a proper inspection warrant or other remedy provided by law to secure entry, the owner, owner's authorized agent or occupant or person having charge, care or control of any building or premises shall not fail or neglect, after proper request is made as herein provided, to promptly permit entry therein by the code official for the purpose of inspection and examination pursuant to this code.

[A] 104.5 Identification. The code official shall carry proper identification when inspecting structures or premises in the performance of duties under this code.

[A] 104.6 Notices and orders. The code official shall issue all necessary notices or orders to ensure compliance with this code.

[A] 104.7 Department records. The code official shall keep official records of applications received, permits and certificates issued, fees collected, reports of inspections, and notices and orders issued. Such records shall be retained in the official records for the period required for retention of public records.

SECTION 105
APPROVAL

[A] 105.1 Modifications. Where there are practical difficulties involved in carrying out the provisions of this code, the code official shall have the authority to grant modifications for individual cases, upon application of the owner or owner's authorized agent, provided that the code official shall first find that special individual reason makes the strict letter of this code impractical, the modification is in conformity with the intent and purpose of this code and such modification does not lessen health and fire- and life-safety requirements. The details of action granting modifications shall be recorded and entered in the files of the Private Sewage Disposal Inspection Department.

[A] 105.2 Alternative materials, design and methods of construction and equipment. The provisions of this code are not intended to prevent the installation of any material or to prohibit any design or method of construction not specifically prescribed by this code, provided that any such alternative has been approved. An alternative material, design or method of construction shall be approved where the code official finds that the proposed design is satisfactory and complies with the intent of the provisions of this code, and that the material, method or work offered is, for the purpose intended, not less than the equivalent of that prescribed in this code in quality, strength, effectiveness, fire resistance, durability and safety. Where the alternative material, design or method of construction is not approved, the code official shall respond in writing, stating the reasons why the alternative was not approved.

[A] 105.2.1 Research reports. Supporting data, where necessary to assist in the approval of materials or assemblies not specifically provided for in this code, shall consist of valid research reports from approved sources.

[A] 105.3 Required testing. Where there is insufficient evidence of compliance with the provisions of this code, or evidence that a material or method does not conform to the requirements of this code, or in order to substantiate claims for alternate materials or methods, the code official shall have the authority to require testing as evidence of compliance at no expense to the jurisdiction.

[A] 105.3.1 Test methods. Test methods shall be as specified in this code or by other recognized test standards. In the absence of recognized and accepted test methods, the code official shall approve the testing procedures.

[A] 105.3.2 Testing agency. Tests shall be performed by an approved agency.

[A] 105.3.3 Test reports. Reports of tests shall be retained by the code official for the period required for retention of public records.

[A] 105.4 Used material and equipment. Materials that are reused shall comply with the requirements of this code for new materials. Materials, equipment and devices shall not be reused unless such elements have been reconditioned, tested and placed in good and proper working condition and approved by the code official.

[A] 105.5 Approved materials and equipment. Materials, equipment and devices approved by the code official shall be constructed and installed in accordance with such approval.

SECTION 106 PERMITS

[A] 106.1 Where required. Work on a *private sewage disposal system* shall not commence until a permit for such work has been issued by the code official.

[A] 106.1.1 Annual permit. Instead of an individual construction permit for each alteration to an already approved system or equipment or appliance installation, the code official is authorized to issue an annual permit upon application therefor to any person, firm or corporation regularly employing one or more qualified tradespersons in the building, structure or on the premises owned or operated by the applicant for the permit.

[A] 106.1.2 Annual permit records. The person to whom an annual permit is issued shall keep a detailed record of alterations made under such annual permit. The code official shall have access to such records at all times or such records shall be filed with the code official as designated.

[A] 106.2 Application for permit. Each application for a permit, with the required fee, shall be filed with the code official on a form furnished for that purpose and shall contain a general description of the proposed work and its location. The application shall contain a description of the type of system, the system location, the occupancy of all parts of the structure and all portions of the site or lot not covered by the structure, and such additional information as is required by the code official. The maximum number of bedrooms for residential occupancies shall be indicated.

[A] 106.2.1 Construction documents. An application for a permit shall be accompanied by not less than two copies of construction documents drawn to scale, with sufficient clarity and detail dimensions showing the nature and character of the work to be performed. Specifications shall include pumps and controls, dose volume, elevation differences (vertical lift), pipe friction loss, pump performance curve, pump model and pump manufacturer. The code official is permitted to waive the requirements for filing construction documents where the work involved is of a minor nature. Where the quality of the materials is essential for conformity to this code, specific information shall be given to establish such quality, and this code shall not be cited, or the term "legal" or its equivalent used as a substitute for specific information.

[A] 106.2.2 Preliminary inspection. Before a permit is issued, the code official is authorized to inspect and evaluate the systems, equipment, buildings, devices, premises and spaces or areas to be used.

[A] 106.2.3 Time limitation of application. An application for a permit for any proposed work shall be deemed to have been abandoned 180 days after the date of filing, unless such application has been pursued in good faith or a permit has been issued; except that the code official shall have the authority to grant one or more extensions of time for additional periods not exceeding 180 days each. The extension shall be requested in writing and justifiable cause demonstrated.

[A] 106.2.4 Previous approvals. This code shall not require changes in the construction documents, construction or designated occupancy of a structure for which a lawful permit has been heretofore issued or otherwise lawfully authorized, and the construction of which has been pursued in good faith within 180 days after the effective date of this code and has not been abandoned.

[A] 106.2.5 Soil data. Soil test reports shall be submitted indicating *soil boring* and percolation test data related to the undisturbed and finished grade elevations, vertical elevation reference point and horizontal reference point. Surface elevations shall be given for all *soil borings*. Soil reports shall bear the signature of a soil tester.

[A] 106.2.6 Site plan. A site plan shall be filed showing to scale the location of all septic tanks, holding tanks or other treatment tanks; building sewers; wells; water mains; water service; streams and lakes; *flood hazard areas*; dosing or pumping chambers; distribution boxes; effluent systems; dual disposal systems; replacement system areas; and the location of all buildings or structures. Separating distances and dimensions shall be shown, including any distance to adjoining property. A vertical elevation reference point and a horizontal reference point shall be indicated. For other than single-family dwellings, grade slope with contours shall be shown for the grade elevation of the entire area of the soil absorption system and the area on all sides for a distance of 25 feet (7620 mm).

[A] 106.3 Permit issuance. The application, construction documents and other data filed by an applicant for permit shall be reviewed by the code official. If the code official finds that the proposed work conforms to the requirements of this code and all laws and ordinances applicable thereto, and that the fees specified in Section 106.4 have been paid, a permit shall be issued to the applicant. A *private sewage disposal system* permit shall not be transferable.

[A] 106.3.1 Approved construction documents. When the code official issues the permit where construction documents are required, the construction documents shall be endorsed in writing and stamped "APPROVED." Such approved construction documents shall not be changed, modified or altered without authorization from the code official. Work shall be done in accordance with the approved construction documents.

The code official shall have the authority to issue a permit for the construction of a part of a *private sewage disposal system* before the construction documents for the whole system have been submitted or approved, provided adequate information and detailed statements have been filed complying with all pertinent requirements of this code. The holder of such permit shall proceed at his or her own risk without assurance that the permit for the entire system will be granted.

[A] 106.3.2 Validity. The issuance of a permit or approval of construction documents shall not be construed to be a permit for, or an approval of, any violation of any of the provisions of this code or of other ordinances of the jurisdiction. No permit presuming to give authority to violate or cancel the provisions of this code shall be valid.

The issuance of a permit based on construction documents and other data shall not prevent the code official from thereafter requiring the correction of errors in said construction documents and other data or from preventing building operations being carried on thereunder when in violation of this code or of other ordinances of the jurisdiction.

[A] 106.3.3 Expiration. Every permit issued by the code official under the provisions of this code shall expire by limitation and become null and void if the work authorized by such permit is not commenced within 180 days from the date of the permit, or if the work authorized by such permit is suspended or abandoned at any time after the work is commenced for a period of 180 days. Before such work can be recommenced, a new permit shall first be obtained and the fee therefor shall be one-half the amount required for a new permit for such work, provided that no changes have been or will be made in the original construction documents for such work, and provided further that such suspension or abandonment has not exceeded one year.

[A] 106.3.4 Extensions. Any permittee holding an unexpired permit shall have the right to apply for an extension of the time within which the permittee will commence work under that permit where work cannot be commenced within the time required by this section for good and satisfactory reasons. The code official shall extend the time for action by the permittee for a period not exceeding 180 days if there is reasonable cause. No permit shall be extended more than once. The fee for an extension shall be one-half the amount required for a new permit for such work.

[A] 106.3.5 Suspension or revocation of permit. The code official shall have the authority to suspend or revoke a permit issued under the provisions of this code wherever the permit is issued in error or on the basis of incorrect, inaccurate or incomplete information, or in violation of any ordinance or regulation or any of the provisions of this code.

[A] 106.3.6 Retention of construction documents. One set of approved construction documents shall be retained by the code official for a period of not less than 180 days from date of completion of the permitted work, or as required by state or local laws. One set of approved construction documents shall be returned to the applicant, and said set shall be kept on the site of the building or work at all times during which the work authorized thereby is in progress.

[A] 106.3.7 Posting of permit. The permit or a copy shall be kept on the site of the work until the completion of the project.

[A] 106.4 Fees. A permit shall not be issued until the fees prescribed in Section 106.4.2 have been paid, and an amendment to a permit shall not be released until the additional fee, if any, due to an increase of the *private sewage disposal system*, has been paid.

[A] 106.4.1 Work commencing before permit issuance. Any person who commences any work on a *private sewage disposal system* before obtaining the necessary permits shall be subject to 100 percent of the usual permit fee in addition to the required permit fees.

[A] 106.4.2 Fee schedule. The fees for all private sewage disposal work shall be as indicated in the following schedule:

[JURISDICTION TO INSERT APPROPRIATE SCHEDULE].

[A] 106.4.3 Fee refunds. The code official shall authorize the refunding of fees as follows:

1. The full amount of any fee paid hereunder that was erroneously paid or collected.
2. Not more than **[SPECIFY PERCENTAGE]** percent of the permit fee paid where no work has been done under a permit issued in accordance with this code.
3. Not more than **[SPECIFY PERCENTAGE]** percent of the plan review fee paid where an application for a permit for which a plan review fee has been paid is withdrawn or canceled before any plan review effort has been expended.

The code official shall not authorize the refunding of any fee paid except upon written application filed by the original permittee no later than 180 days after the date of fee payment.

SECTION 107 INSPECTIONS

[A] 107.1 Required inspections. After issuing a permit, the code official shall conduct inspections from time to time during and upon completion of the work for which a permit has been issued. A record of all such examinations and inspections and of all violations of this code shall be maintained by the code official.

[A] 107.1.1 Concealed work. It shall be the duty of the permit applicant to cause the work to remain visible and able to be accessed for inspection purposes. Neither the code official nor the jurisdiction shall be liable for expense entailed in the removal or replacement of any material required to allow inspection.

[A] 107.1.2 Other inspections. The code official is authorized to make or require other inspections to ascertain compliance with the provisions of this code and other laws that are enforced by the department.

[A] 107.1.3 Approved inspection agencies. The code official shall accept reports of approved inspection agencies provided that such agencies satisfy the requirements as to qualifications and reliability.

[A] 107.2 Special inspections. Special inspections of alternative engineered design *private sewage disposal systems* shall be conducted in accordance with Sections 107.2.1 and 107.2.2.

[A] 107.2.1 Periodic inspection. The registered design professional or designated inspector shall periodically inspect and observe the alternative engineered design to determine that the installation is in accordance with the approved plans. Discrepancies shall be brought to the immediate attention of the *private sewage disposal system* contractor for correction. Records shall be kept of all inspections.

[A] 107.2.2 Written report. The registered design professional shall submit a final report in writing to the code official upon completion of the installation, certifying that the alternative engineered design conforms to the approved construction documents. A notice of approval for the *private sewage disposal system* shall not be issued until a written certification has been submitted.

[A] 107.3 Contractor's responsibilities. It shall be the duty of every contractor who enters into contracts for the installation or repair of *private sewage disposal systems* for which a permit is required to comply with adopted state and local rules and regulations concerning licensing.

[A] 107.3.1 Inspection requests. It shall be the duty of the holder of the permit or their duly authorized agent to notify the code official when work is ready for inspection. It shall be the duty of the permit holder to provide access to and means for inspections of such work that is required by this code.

[A] 107.4 Approval required. Work shall not be done beyond the point indicated in each successive inspection without first obtaining the approval of the code official. The code official, upon notification, shall make the requested inspections and shall either indicate the portion of the construction that is satisfactory as completed, or notify the permit holder or his or her agent wherein the same fails to comply with this code. Any portions that do not comply shall be corrected and such portion shall not be covered or concealed until authorized by the code official.

[A] 107.5 Evaluation and follow-up inspection services. Prior to the approval of a prefabricated construction assembly having concealed work and the issuance of a permit, the code official shall require the submittal of an evaluation report on each prefabricated construction assembly, indicating the complete details of the *private sewage disposal system*, including a description of the system and its components, the basis on which the system is being evaluated, test results and similar information and other data as necessary for the code official to determine conformance to this code.

[A] 107.5.1 Evaluation service. The code official shall designate the evaluation service of an approved agency as the evaluation agency, and review such agency's evaluation report for adequacy and conformance to this code.

[A] 107.5.2 Follow-up inspection. Except where ready access is provided to *private sewage disposal systems*, service equipment and accessories for complete inspection at the site without disassembly or dismantling, the code official shall conduct the in-plant inspections as frequently as necessary to ensure conformance to the approved evaluation report or shall designate an independent, approved inspection agency to conduct such inspections. The inspection agency shall furnish the code official with the follow-up inspection manual and a report of inspections on request, and the installation shall have an identifying label permanently affixed to the system indicating that factory inspections have been performed.

[A] 107.5.3 Test and inspection records. Required test and inspection records shall be available to the code official at all times during the fabrication of the installation and the erection of the building; or such records as the code official designates shall be filed.

[A] 107.6 Testing. Installations shall be tested as required in this code and in accordance with Sections 107.6.1 through 107.6.3. Tests shall be made by the permit holder and observed by the code official.

[A] 107.6.1 New, altered, extended or repaired installations. New installations and parts of existing installations that have been altered, extended, renovated or repaired shall be tested as prescribed herein to disclose leaks and defects.

[A] 107.6.2 Apparatus, instruments, material and labor for tests. Apparatus, instruments, material and labor required for testing an installation or part thereof shall be furnished by the permit holder.

[A] 107.6.3 Reinspection and testing. Where any work or installation does not pass an initial test or inspection, the necessary corrections shall be made so as to achieve compliance with this code. The work or installation shall then be resubmitted to the code official for inspection and testing.

[A] 107.7 Approval. After the prescribed inspections indicate that the work complies in all respects with this code, a notice of approval shall be issued by the code official.

[A] 107.7.1 Revocation. The code official is authorized to, in writing, suspend or revoke a notice of approval issued under the provisions of this code wherever the notice is issued in error, on the basis of incorrect information supplied, or where it is determined that the building or structure, premise or portion thereof is in violation of any ordinance or regulation or any of the provisions of this code.

[A] 107.8 Temporary connection. The code official shall have the authority to allow the temporary connection of an installation to the sources of energy for the purpose of testing the installation or for use under a temporary certificate of occupancy.

[A] 107.9 Connection of service utilities. No person shall make connections from a utility, source of energy, fuel or power to any building or system that is regulated by this code for which a permit is required until authorized by the code official.

SECTION 108 VIOLATIONS

[A] 108.1 Unlawful acts. It shall be unlawful for any person, firm or corporation to erect, construct, alter, repair, remove, demolish or use any *private sewage disposal system*, or cause same to be done, in conflict with or in violation of any of the provisions of this code.

[A] 108.2 Notice of violation. The code official shall serve a notice of violation or order to the person responsible for the erection, installation, alteration, extension, repair, removal or demolition of private sewage disposal work in violation of the provisions of this code; in violation of a detailed statement or the approved construction documents thereunder or in violation of a permit or certificate issued under the provisions of this code. Such order shall direct the discontinuance of the illegal action or condition and the abatement of the violation.

[A] 108.3 Prosecution of violation. If the notice of violation is not complied with promptly, the code official shall request the legal counsel of the jurisdiction to institute the appropriate proceeding at law or in equity to restrain, correct or abate such violation, or to require the removal or termination of the unlawful system in violation of the provisions of this code or of the order or direction made pursuant thereto.

[A] 108.4 Violation penalties. Any person who shall violate a provision of this code or fail to comply with any of the requirements thereof or who shall erect, install, alter or repair private sewage disposal work in violation of the approved construction documents or directive of the code official, or of a permit or certificate issued under the provisions of this code, shall be guilty of a [SPECIFY OFFENSE], punishable by a fine of not more than [AMOUNT] dollars or by imprisonment not exceeding [NUMBER OF DAYS], or both such fine and imprisonment. Each day that a violation continues after due notice has been served shall be deemed a separate offense.

[A] 108.5 Stop work orders. Upon notice from the code official, work on any *private sewage disposal system* that is being performed contrary to the provisions of this code or in a dangerous or unsafe manner shall immediately cease. Such notice shall be in writing and shall be given to the owner of the property, to the owner's authorized agent or to the person performing the work. The notice shall state the conditions under which work is authorized to resume. Where an emergency exists, the code official shall not be required to give a written notice prior to stopping the work. Any person who shall continue any work on the system after having been served with a stop work order, except such work as that person is directed to perform to remove a violation or unsafe condition, shall be liable to a fine of not less than [AMOUNT] dollars or more than [AMOUNT] dollars.

[A] 108.6 Abatement of violation. The imposition of the penalties herein prescribed shall not preclude the legal officer of the jurisdiction from instituting appropriate action to prevent unlawful construction or to restrain, correct or abate a violation; to prevent illegal occupancy of a building, structure or premises or to stop an illegal act, conduct, business or use of the *private sewage disposal system* on or about any premises.

[A] 108.7 Unsafe systems. Any *private sewage disposal system* regulated by this code that is unsafe or constitutes a health hazard, insanitary condition or is otherwise dangerous to human life is hereby declared unsafe. Any use of *private sewage disposal systems* regulated by this code constituting a hazard to safety, health or public welfare by reason of inadequate maintenance, dilapidation, obsolescence, disaster, damage or abandonment is hereby declared an unsafe use. Any such unsafe equipment is hereby declared to be a public *nuisance* and shall be abated by repair, rehabilitation, demolition or removal.

[A] 108.7.1 Authority to condemn equipment. Where the code official determines that any *private sewage disposal system*, or portion thereof, regulated by this code has become hazardous to life, health or property or has become insanitary, the code official shall order in writing that such system be either removed or restored to a safe or sanitary condition. A time limit for compliance with such order shall be specified in the written notice. A defective *private sewage disposal system* shall not be used or maintained after receiving such notice. Where such system is to be disconnected, written notice as prescribed in Section 108.2 shall be given. In cases of immediate danger to life or property, such disconnection shall be made immediately without such notice.

[A] 108.7.2 Authority to disconnect service utilities. The code official shall have the authority to authorize disconnection of utility service to the building, structure or system regulated by the technical codes in case of emergency, where necessary, to eliminate an immediate danger to life or property. Where possible, the owner, the owners's authorized agent and occupant of the building, structure or service system shall be notified of the decision to disconnect utility service prior to taking such action. If not notified prior to disconnecting, the owner or occupant of the building, structure or service systems shall be notified in writing as soon as is practical thereafter.

SECTION 109 MEANS OF APPEAL

[A] 109.1 Application for appeal. Any person shall have the right to appeal a decision of the code official to the board of appeals. An application for appeal shall be based on a claim that the true intent of this code or the rules legally adopted thereunder has been incorrectly interpreted, the provisions of this code do not fully apply or an equally good or better form of construction is proposed. The application shall be filed on a form obtained from the code official within 20 days after the notice was served.

[A] 109.2 Membership of board. The board of appeals shall consist of five members appointed by the chief appointing authority as follows: one for 5 years, one for 4 years, one for

3 years, one for 2 years and one for 1 year. Thereafter, each new member shall serve for 5 years or until a successor has been appointed.

[A] 109.2.1 Qualifications. The board of appeals shall consist of five individuals, one from each of the following professions or disciplines.

1. Registered design professional who is a registered architect; or a builder or superintendent of building construction with not less than 10 years' experience, 5 years of which shall have been in responsible charge of work.
2. Registered design professional with structural engineering or architectural experience.
3. Registered design professional with mechanical and plumbing engineering experience; or a mechanical and plumbing contractor with not less than 10 years' experience, 5 years of which shall have been in responsible charge of work.
4. Registered design professional with electrical engineering experience; or an electrical contractor with not less than 10 years' experience, 5 years of which shall have been in responsible charge of work.
5. Registered design professional with fire protection engineering experience; or a fire protection contractor with not less than 10 years' experience, 5 years of which shall have been in responsible charge of work.

[A] 109.2.2 Alternate members. The chief appointing authority shall appoint two alternate members who shall be called by the board chairman to hear appeals during the absence or disqualification of a member. Alternate members shall possess the qualifications required for board membership, and shall be appointed for 5 years or until a successor has been appointed.

[A] 109.2.3 Chairman. The board shall annually select one of its members to serve as chairman.

[A] 109.2.4 Disqualification of a member. A member shall not hear an appeal in which that member has any personal, professional or financial interest.

[A] 109.2.5 Secretary. The chief administrative officer shall designate a qualified clerk to serve as secretary to the board. The secretary shall file a detailed record of all proceedings in the office of the chief administrative officer.

[A] 109.2.6 Compensation of members. Compensation of members shall be determined by law.

[A] 109.3 Notice of meeting. The board shall meet upon notice from the chairman, within 10 days of the filing of an appeal or at stated periodic meetings.

[A] 109.4 Open hearing. Hearings before the board shall be open to the public. The appellant, the appellant's representative, the code official and any person whose interests are affected shall be given an opportunity to be heard.

[A] 109.4.1 Procedure. The board shall adopt and make available to the public through the secretary procedures under which a hearing will be conducted. The procedures shall not require compliance with strict rules of evidence, but shall mandate that only relevant information be received.

[A] 109.5 Postponed hearing. When five members are not present to hear an appeal, either the appellant or the appellant's representative shall have the right to request a postponement of the hearing.

[A] 109.6 Board decision. The board shall modify or reverse the decision of the code official by a concurring vote of three members.

[A] 109.6.1 Resolution. The decision of the board shall be by resolution. Certified copies shall be furnished to the appellant and to the code official.

[A] 109.6.2 Administration. The code official shall take immediate action in accordance with the decision of the board.

[A] 109.7 Court review. Any person, whether or not a previous party of the appeal, shall have the right to apply to the appropriate court for a writ of certiorari to correct errors of law. Application for review shall be made in the manner and time required by law following the filing of the decision in the office of the chief administrative officer.

SECTION 110
TEMPORARY EQUIPMENT, SYSTEMS AND USES

[A] 110.1 General. The code official is authorized to issue a permit for temporary equipment, systems and uses. Such permits shall be limited as to time of service, but shall not be permitted for more than 180 days. The code official is authorized to grant extensions for demonstrated cause.

[A] 110.2 Conformance. Temporary equipment, systems and uses shall conform to the structural strength, fire safety, means of egress, accessibility, light, ventilation and sanitary requirements of this code as necessary to ensure the public health, safety and general welfare.

[A] 110.3 Temporary utilities. The code official is authorized to give permission to temporarily supply utilities before an installation has been fully completed and the final certificate of completion has been issued. The part covered by the temporary certificate shall comply with the requirements specified for temporary lighting, heat or power in the code.

[A] 110.4 Termination of approval. The code official is authorized to terminate such permit for a temporary structure or use and to order the temporary structure or use to be discontinued.

CHAPTER 2

DEFINITIONS

User note:

About this chapter: *Codes, by their very nature, are technical documents. Every word, term and punctuation mark can add to or change the meaning of a technical requirement. It is necessary to maintain a consensus on the specific meaning of each term contained in the code. Chapter 2 performs this function by stating clearly what specific terms mean for the purpose of the code.*

SECTION 201 GENERAL

201.1 Scope. Unless otherwise expressly stated, the following words and terms shall, for the purposes of this code, have the meanings indicated in this chapter.

201.2 Interchangeability. Words used in the present tense include the future; words in the masculine gender include the feminine and neuter; the singular number includes the plural and the plural, the singular.

201.3 Terms defined in other codes. Where terms are not defined in this code and are defined in the *International Building Code* or the *International Plumbing Code*, such terms shall have meanings ascribed to them as in those codes.

201.4 Terms not defined. Where terms are not defined through the methods authorized by this section, such terms shall have ordinarily accepted meanings such as the context implies.

SECTION 202 GENERAL DEFINITIONS

AGGREGATE. Graded hard rock that has been washed with water under pressure over a screen during or after grading to remove fine material and with a hardness value of 3 or greater on Mohs' Scale of Hardness. Aggregate that will scratch a copper penny without leaving any residual rock material on the coin has a hardness value of 3 or greater on Mohs' Scale of Hardness.

[P] AIR BREAK (Drainage System). A piping arrangement in which a drain from a fixture, appliance or device discharges indirectly into another fixture, receptacle or interceptor at a point below the flood level rim and above the trap seal.

ALLUVIUM. Soil deposited by floodwaters.

BEDROCK. The rock that underlies soil material or is located at the earth's surface. Bedrock is encountered when the weathered in-place consolidated material, larger than 0.08 inch (2 mm) in size, is more than 50 percent by volume.

CESSPOOL. A covered excavation in the ground receiving sewage or other organic wastes from a drainage system that is designed to retain the organic matter and solids, permitting the liquids to seep into the soil cavities.

CLEAR-WATER WASTES. Cooling water and condensate drainage from refrigeration compressors and air-conditioning equipment, water used for equipment chilling purposes, liquid having no impurities or where impurities have been reduced below a minimum concentration considered harmful, and cooled condensate from steam-heating systems or other equipment.

[A] CODE OFFICIAL. The officer or other designated authority charged with administration and enforcement of this code or a duly authorized representative.

COLLUVIUM. Soil transported under the influence of gravity.

COLOR. The moist color of the soil based on Munsell soil color charts.

[A] CONSTRUCTION DOCUMENTS. All the written, graphic and pictorial documents prepared or assembled for describing the design, location and physical characteristics of the elements of the project necessary for obtaining a building permit. The construction drawings shall be drawn to an appropriate scale.

CONVENTIONAL SOIL ABSORPTION SYSTEM. A system employing gravity flow from the septic or other treatment tank and applying effluent to the soil through the use of a *seepage trench*, bed or pit.

[BS] DESIGN FLOOD ELEVATION. The elevation of the "design flood," including wave height, relative to the datum specified on the community's legally designated flood hazard map. In areas designated as Zone AO, the *design flood elevation* shall be the elevation of the highest existing grade of the *building's* perimeter plus the depth number (in feet) specified on the flood hazard map. In areas designated as Zone AO where a depth number is not specified on the map, the depth number shall be taken as being equal to 2 feet (610 mm).

DETAILED SOIL MAP. A map prepared by or for a state or federal agency participating in the National Cooperative Soil Survey showing soil series, type and phases at a scale of not more than 2,000 feet to the inch (24 m/mm) and which includes related explanatory information.

DOSING SOIL ABSORPTION SYSTEM. A system employing a pump or automatic siphon to elevate or distribute effluent to the soil through the use of a *seepage trench* or bed.

EFFLUENT. Liquid discharged from a septic or other treatment tank.

[BS] FLOOD HAZARD AREA. The greater of the following two areas:

1. The area within a flood plain subject to a 1-percent or greater chance of flooding in any given year.
2. The area designated as a flood hazard area on a community's flood hazard map or as otherwise legally designated.

HIGH GROUND WATER. Soil saturation zones, including perched water tables, shallow regional ground water tables or aquifers, or zones seasonally, periodically or permanently saturated.

HOLDING TANK. An approved water-tight receptacle for collecting and holding sewage.

HORIZONTAL REFERENCE POINT. A stationary, easily identifiable point to which horizontal dimensions are related.

LEGAL DESCRIPTION. An accurate metes and bounds description, a lot and block number in a recorded subdivision, a recorded assessor's plat or a public land survey description to the nearest 40 acres (16 ha).

MANHOLE. An opening of sufficient size to permit a person to gain access to a sewer or any portion of a *private sewage disposal system.*

MOBILE UNIT. A structure of vehicular, portable design, built on a chassis and designed to be moved from one site to another and to be used with or without a permanent foundation.

MOBILE UNIT PARK. Any plot or plots of ground owned by a person, state or local government upon which two or more units, occupied for dwelling or sleeping purposes regardless of mobile unit ownership, are located and whether or not a charge is made for such accommodation.

[P] NUISANCE. Public nuisance as known in common law or equity jurisprudence; whatever is dangerous to human life or detrimental to health; whatever building, structure or premises is not sufficiently ventilated, sewered, drained, cleaned or lighted, in reference to its intended use; and whatever renders the air, human food, drink or water supply unwholesome.

PAN. A soil horizon cemented with any one of a number of cementing agents such as iron, organic matter, silica, calcium, carbonate, gypsum or a combination of chemicals. Pans will resist penetration from a knife blade and are slowly permeable horizons or are impermeable.

PERCOLATION TEST. The method of testing absorption qualities of the soil (see Section 404).

PERMEABILITY. The ease with which liquids move through the soil. One of the soil qualities listed in soil survey reports.

PRESSURE DISTRIBUTION SYSTEM. A soil absorption system using a pump or automatic siphon and smaller diameter distribution piping with small-diameter perforations to introduce effluent into the soil.

PRIVATE SEWAGE DISPOSAL SYSTEM. A sewage treatment and disposal system serving a single structure with a septic tank and soil absorption field located on the same parcel as the structure. This term also means an alternative sewage disposal system, including a substitute for the septic tank or soil absorption field, a holding tank, a system serving more than one structure or a system located on a different parcel than the structure. A private sewage disposal system is permitted to be owned by the property owner or a special-purpose district.

PRIVY. A structure, not connected to a plumbing system, that is used by persons for the deposition of human body waste.

[A] REGISTERED DESIGN PROFESSIONAL. An individual who is registered or licensed to practice their respective design profession, as defined by the statutory requirements of the professional registration laws of the state or jurisdiction in which the project is to be constructed.

SEEPAGE BED. An excavated area more than 5 feet (1524 mm) wide that contains a bedding of aggregate and has more than one distribution line.

SEEPAGE PIT. An underground receptacle constructed to permit disposal of effluent or clear wastes by soil absorption through its floor and walls.

SEEPAGE TRENCH. An area excavated 1 foot to 5 feet (305 mm to 1524 mm) wide containing a bedding of aggregate and a single distribution line.

SEPTAGE. All sludge, scum, liquid and any other material removed from a private sewage treatment and disposal system.

SEPTIC TANK. A tank that receives and partially treats sewage through processes of sedimentation, flotation and bacterial action to separate solids from the liquid in the sewage, and which discharges the liquid to a soil absorption system.

SOIL. The unconsolidated material over bedrock, 0.08 inch (2 mm) and smaller.

SOIL BORING. An observation pit dug by hand or backhoe, a hole dug by augering or a soil core taken intact and undisturbed with a probe.

SOIL MOTTLES. Spots, streaks or contrasting soil colors usually caused by soil saturation for one period of a normal year, with a color value of 4 or more and a chroma of 2 or less. Gray-colored mottles are called low chroma; reddish-brown, red- and yellow-colored mottles are called high chroma.

SOIL SATURATION. The state in which all pores in a soil are filled with water. Water will flow from saturated soil into a bore hole.

VENT CAP. An approved appurtenance used for covering the vent terminal of an effluent disposal system to avoid closure by mischief or debris and still permit circulation of air within the system.

VERTICAL ELEVATION REFERENCE POINT. An easily identifiable stationary point or object of constant elevation for establishing the relative elevation of percolation tests, *soil borings* and other locations.

WATERCOURSE. A stream usually flowing in a particular direction, though it need not flow continually and is sometimes dry. A *watercourse* flows in a definite channel, with a bed, sides or banks, and usually discharges itself into some other stream or body of water. It must be something more than mere surface drainage over the entire face of a tract of land, occasioned by unusual freshets or other extraordinary cause. It does not include the water flowing in the hollows or ravines in land, which is the mere surface water from rains or melting snows, and is discharged through them from a higher to a lower level, but which at other times are destitute of water. Such hollows or ravines are not, in legal contemplation, *watercourses.*

WORKMANSHIP. Work of such character that will fully secure the results sought in all the sections of this code as intended for the health, safety and welfare protection of all individuals.

CHAPTER 3

GENERAL REGULATIONS

User note:

About this chapter: *Chapter 3 covers general regulations for private sewage disposal installations. As many of these requirements would need to be repeated in Chapters 3 through 13, placing such requirements in only one location eliminates code development coordination issues associated with the same requirement in multiple locations. These general requirements can be superseded by more specific requirements for certain applications in Chapters 3 through 13.*

SECTION 301
GENERAL

301.1 Scope. The provisions of this chapter shall govern the general regulations of *private sewage disposal systems*, including specific limitations and *flood hazard areas*.

SECTION 302
SPECIFIC LIMITATIONS

302.1 Domestic waste. Waste and sewage derived from ordinary living uses shall enter the septic or treatment tank unless otherwise specifically exempted by the code official or this code.

302.2 Cesspools and privies. Privies shall be prohibited. *Cesspools* shall be prohibited, except where approved by the code official. Where approved, *cesspools* shall be designed and installed in accordance with Chapter 10.

302.3 Industrial wastes. The code official shall approve the method of treatment and disposal of all waste products from manufacturing or industrial operations, including combined industrial and domestic waste.

302.4 Detrimental or dangerous waste. Material such as ashes, cinders or rags; flammable, poisonous or explosive liquids or gases; oil, grease or other insoluble material that is capable of obstructing, damaging or overloading the *private sewage disposal system*, or is capable of interfering with the normal operation of the *private sewage disposal system*, shall not be deposited, by any means, into such systems. The code official shall approve the method of treatment and disposal.

302.5 Clear water. The discharge of surface, rain or other clear water into a *private sewage disposal system* shall be prohibited.

302.6 Water softener and iron filter backwash. Water softener or iron filter discharge shall be indirectly connected by means of an air gap to the *private sewage disposal system* or discharge onto the ground surface, provided that a *nuisance* is not created.

302.7 Food waste disposals. Where a food waste disposal connects to a *private sewage disposal system*, the system shall be designed to accommodate the solids loading from the disposal unit.

SECTION 303
FLOOD HAZARD AREAS

[BS] 303.1 General. Soil absorption systems shall be located outside of *flood hazard areas*.

> **Exception:** Where suitable soil absorption sites outside of the flood hazard area are not available, the soil absorption site is permitted to be located within the flood hazard area. The soil absorption site shall be located to minimize the effects of inundation under conditions of the design flood.

[BS] 303.2 Tanks. In *flood hazard areas*, tanks shall be anchored to counter buoyant forces during condition of the design flood. The vent termination and service manhole of the tank shall be not less than 2 feet (610 mm) above the *design flood elevation* or fitted with covers designed to prevent the inflow of floodwater or outflow of the contents of the tanks during conditions of the design flood.

[BS] 303.3 Mound systems. Mound systems shall be prohibited in *flood hazard areas*.

SECTION 304
ALTERNATIVE ENGINEERED DESIGN

304.1 Alternative engineered design. The design, documentation, inspection, testing and approval of an alternative engineered design *private sewage disposal system* shall comply with Sections 304.1.1 through 304.6.

> **304.1.1 Design criteria.** An alternative engineered design shall conform to the intent of the provisions of this code and shall provide an equivalent level of quality, strength, effectiveness, fire resistance, durability and safety. Material, equipment or components shall be designed and installed in accordance with the manufacturer's instructions.

304.2 Submittal. The registered design professional shall indicate on the permit application that the *private sewage disposal system* is an alternative engineered design. The permit and permanent permit records shall indicate that an alternative engineered design was part of the approved installation.

304.3 Technical data. The registered design professional shall submit sufficient technical data to substantiate the proposed alternative engineered design and to prove that the performance meets the intent of this code.

304.4 Construction documents. The registered design professional shall submit to the code official two complete sets of signed and sealed construction documents for the alternative engineered design.

304.5 Design approval. Where the code official determines that the alternative engineered design conforms to the intent of this code, the *private sewage disposal system* shall be approved. If the alternative engineered design is not approved, the code official shall notify the registered design professional in writing, stating the reasons therefor.

304.6 Inspection and test. The alternative engineered design shall be inspected in accordance with the requirements of Section 107.

CHAPTER 4

SITE EVALUATION AND REQUIREMENTS

User note:

About this chapter: *Disposal systems covered in this code rely on the subsurface soil's abilities to accept the nonpotable water that is discharged by the treatment methods described in the code. Chapter 4 provides the methods for the evaluation of the soil in the planned disposal area.*

SECTION 401
GENERAL

401.1 Scope. The provisions of this chapter shall govern the evaluation of and requirements for *private sewage disposal system* sites.

401.2 Site evaluation. Site evaluation shall include soil conditions, properties and permeability, depth to zones of soil saturation, depth to bedrock, slope, landscape position, all setback requirements and the presence of *flood hazard areas.* Soil test data shall relate to the undisturbed elevations, and a vertical elevation reference point or benchmark shall be established. Evaluation data shall be reported on approved forms. Reports shall be filed within 30 days of the completion of testing for all sites investigated.

401.3 Replacement system area. On each parcel of land being initially developed, sufficient area of suitable soils based on the soil tests and system location and site requirements of this code for one replacement system shall be established. Where bore hole test data in the replacement system area are equivalent to data in the proposed system area, the percolation test is not required.

401.3.1 Nonconforming site conditions. Where site conditions do not permit replacement systems in accordance with this code and an alternative system is used, the alternative system shall be approved in accordance with Section 105.

401.3.2 Undisturbed site. The replacement system shall not be disturbed to the extent that the site area is no longer suitable. The replacement system area shall not be used for construction of buildings, parking lots or parking areas, below-ground swimming pools or any other use that will adversely affect the replacement area.

SECTION 402
SLOPE

402.1 General. A *conventional soil absorption system* shall not be located on land with a slope greater than 20 percent. A *conventional soil absorption system* shall be located not less than 20 feet (6096 mm) from the crown of land with a slope greater than 20 percent, except where the top of the aggregate of a system is at or below the bottom of an adjacent roadside ditch. Where a more restrictive land slope is to be observed for a soil absorption system, other than a *conventional soil absorption system*, the more restrictive land slope specified in the design sections of this code shall apply.

SECTION 403
SOIL BORINGS AND EVALUATION

403.1 Soil borings and profile descriptions. *Soil borings* shall be conducted on all sites, regardless of the type of private sewage system planned to serve the parcel. Borings shall extend not less than 3 feet (914 mm) below the bottom of the proposed system. Borings shall be of sufficient size and extent to determine the soil characteristics important to an onsite liquid waste disposal system. Borehole data shall be used to determine the suitability of soils at the site with respect to zones of seasonal or permanent soil saturation and the depth to bedrock. Borings shall be conducted prior to percolation tests to determine whether the soils are suitable to warrant such tests and, if suitable, at what depth percolation tests shall be conducted. The use of power augers for *soil borings* is prohibited. *Soil borings* shall be conducted and reported in accordance with Sections 403.1.1 through 403.1.5. Where it is not practical to have borings made with a backhoe, such borings shall be augered or dug by hand.

403.1.1 Number. There shall be not less than three borings per soil absorption site. Where necessary, more *soil borings* shall be made for an accurate evaluation of a site. Borings shall be constructed to a depth of not less than 3 feet (914 mm) below the proposed depth of the system.

Exception: On new parcels, the requirement of six borings (three for initial area and three for replacement area) shall be reduced to five where the initial and replacement system areas are contiguous and one boring is made on each outer corner of the contiguous area and the fifth boring is made between the system areas (see Appendix A, Figure A-1).

403.1.2 Location. Each borehole shall be accurately located and referenced to the vertical elevation and horizontal reference points. Reports of boring location shall either be drawn to scale or have the horizontal dimensions clearly indicated between the borings and the horizontal reference point.

403.1.3 Soil description. Soil profile descriptions shall be written for all borings. The thickness in inches (mm) of the different soil horizons observed shall be indicated. Horizons shall be differentiated on the basis of color, texture, soil mottles or bedrock. Depths shall be measured from the ground surface.

403.1.4 Soil mottles. Seasonal or periodic soil saturation zones shall be estimated at the highest level of soil mottles. The code official shall require, where deemed necessary, a detailed description of the soil mottling on a marginal site. The abundance, size, contrast and color of the soil mottles shall be described in the following manner:

Abundance shall be described as "few" if the mottled color occupies less than 2 percent of the exposed surface; "common" if the mottled color occupies from 2 to 20 percent of the exposed surface; or "many" if the mottled color occupies more than 20 percent of the exposed surface. Size refers to length of the mottle measured along the longest dimension and shall be described as "fine" if the mottle is less than 0.196 inch (5 mm); medium if the mottle is from 0.196 inch to 1.590 inches (5 mm to 40 mm); or coarse if the mottle is larger than 1.590 inches (40 mm). Contrast refers to the difference in color between the soil mottle and the background color of the soil and is described as "faint" if the mottle is evident but recognizable with close examination; "distinct" if the mottle is readily seen but not striking; or "prominent" if the mottle is obvious and one of the outstanding features of the horizon. The color(s) of the mottle(s) shall be indicated.

403.1.5 Observed ground water. The depth to ground water, if present, shall be reported. Observed ground water shall be reported at the level that ground water reaches in the soil borehole or the highest level of sidewall seepage into the boring. Measurements shall be made from ground level. Soil located above the water level in the boring shall be checked for the presence of soil mottles.

403.2 Color patterns not indicative of soil saturation. The following soil conditions shall be reported, but shall not be interpreted as color patterns caused by wetness or saturation. Soil profiles with an abrupt textural change with finer-textured soils overlying more than 4 feet (1219 mm) of unmottled, loamy sand or coarser soils can have a mottled zone for the finer textured material. Where the mottled zone is less than 12 inches (305 mm) thick and located immediately above the textural change, a soil absorption system shall be permitted in the loamy sand or coarser material below the mottled layer. The site shall be considered to be unsuitable where any soil mottles occur within the sandy material. The code official shall consider certain coarse sandy loam soils to be included as a coarse material.

403.2.1 Other soil color patterns. Soil mottles occur that are not caused by seasonal or periodic soil saturation zones. Examples of such soil conditions not limited by enumeration are soil mottles formed from residual sandstone deposits; soil mottles formed from uneven weathering of glacially deposited material or glacially deposited material that is naturally gray in color, including any concretionary material in various stages of decomposition; deposits of lime in a profile derived from highly calcareous parent material; light-colored silt coats deposited on soil bed faces; and soil mottles usually vertically oriented along old or decayed root channels with a dark organic stain usually present in the center of the mottled area.

403.2.2 Reporting exceptions. The site evaluator shall report any mottled soil condition. The observation of soil mottles not caused by soil saturation shall be reported. On request, the code official shall make a determination of the acceptability of the site.

403.3 Bedrock. The depth of the bedrock, except sandstone, shall be established at the depth in a *soil boring* where more than 50 percent of the weathered-in-place material is consolidated. Sandstone bedrock shall be established at the depth where an increase in resistance to penetration of a knife blade occurs.

403.4 Alluvial and colluvial deposits. Subsurface soil absorption systems shall not be placed in alluvial and colluvial deposits with shallow depths, extended periods of saturation or possible flooding.

SECTION 404 PERCOLATION OR PERMEABILITY EVALUATION

404.1 General. The permeability of the soil in the proposed absorption system shall be determined by percolation tests or permeability evaluation.

404.2 Percolation tests and procedures. Not less than three percolation tests in each system area shall be conducted. The holes shall be spaced uniformly in relation to the bottom depth of the proposed absorption system. More percolation tests shall be made where necessary, depending on system design.

404.2.1 Percolation test hole. The test hole shall be dug or bored. The test hole shall have vertical sides and a horizontal dimension of 4 inches to 8 inches (102 mm to 203 mm). The bottom and sides of the hole shall be scratched with a sharp-pointed instrument to expose the natural soil. Loose material shall be removed from the hole, and the bottom shall be covered with 2 inches (51 mm) of gravel or coarse sand.

404.2.2 Test procedure, sandy soils. The hole shall be filled with clear water to not less than 12 inches (305 mm) above the bottom of the hole for tests in sandy soils. The time for this amount of water to seep away shall be determined and this procedure shall be repeated if the water from the second filling of the hole seeps away in 10 minutes or less. The test shall proceed as follows: Water shall be added to a point not more than 6 inches (152 mm) above the gravel or coarse sand. Thereupon, from a fixed reference point, water levels shall be measured at 10-minute intervals for a period of 1 hour. Where 6 inches (152 mm) of water seeps away in less than 10 minutes, a shorter interval between measurements shall be used, but the water depth shall not exceed 6 inches (152 mm) in any case. Where 6 inches (152 mm) of water seeps away in less than 2 minutes, the test shall be stopped and a rate of less than 3 minutes per inch (7.2 s/mm) shall be reported. The final water level drop shall be used to calculate the percolation rate. Soils not meeting the requirements of this section shall be tested in accordance with Section 404.2.3.

404.2.3 Test procedure, other soils. The hole shall be filled with clear water, and a minimum water depth of 12 inches (305 mm) shall be maintained above the bottom of the hole for a 4-hour period by refilling whenever necessary or by use of an automatic siphon. Water remaining in the hole after 4 hours shall not be removed. Thereafter, the soil shall be allowed to swell not less than 16 hours or more than 30 hours. Immediately after the soil swelling period, the measurements for determining the percolation rate shall be made as follows: Any soil sloughed into the hole shall be removed, and the water level shall be adjusted to 6 inches (152 mm) above the gravel or coarse sand. Thereupon, from a fixed reference point, the water level shall be measured at 30-minute intervals for a period of 4 hours, unless two successive water level drops do not vary by more than $^1/_{16}$ inch (1.59 mm). Not less than three water level drops shall be observed and recorded. The hole shall be filled with clear water to a point not more than 6 inches (152 mm) above the gravel or coarse sand whenever it becomes nearly empty. The water level shall not be adjusted during the three measurement periods except to the limits of the last measured water level drop. Where the first 6 inches (152 mm) of water seeps away in less than 30 minutes, the test shall be performed again for a period of 1 hour with measurements performed every 10 minutes. The water depth shall not exceed 5 inches (127 mm) at any time during the measurement period. The drop that occurs during the final measurement period shall be used in calculating the percolation rate.

404.2.4 Mechanical test equipment. Mechanical percolation test equipment shall be of an approved type.

404.3 Permeability evaluation. Soil shall be evaluated for estimated percolation based on structure and texture in accordance with accepted soil evaluation practices. Borings shall be made in accordance with Section 404.2 for evaluating the soil.

SECTION 405
SOIL VERIFICATION

405.1 Verification. Where required by the code official, depth to soil mottles, depth to high ground water, soil textures, depth to bedrock and land slope shall be verified by the code official. The code official shall require, where necessary, backhoe pits to be provided for verification of *soil boring* data. Where required by the code official, the results of percolation tests or permeability evaluation shall be subject to verification. The code official shall require, where necessary, that percolation tests be conducted under supervision. Where the natural soil condition has been altered by filling or other methods used to improve wet areas, the code official shall require, where necessary, observation of high ground water levels under saturated soil conditions. Detailed soil maps, or other adequate information, shall be used for determining estimated percolation rates and other soil characteristics.

405.2 Monitoring ground water levels. A property owner or developer shall have the option to provide documentation that soil mottling or other color patterns at a particular site are not an indication of seasonally saturated soil conditions of high ground water levels. Direct observation shall be used to document ground water levels. Monitoring shall be in accordance with the procedures cited in Sections 405.2.1 through 405.2.6.

405.2.1 Precipitation. Monitoring shall be performed at a time of the year when maximum ground water elevation occurs. In determining whether a near-normal season has occurred where sites are subject to broad regional water tables, such as large areas of sandy soils, the fluctuation over the several-year cycle shall be considered. In such cases, data obtained from the United States Geological Survey (USGS) shall be used to determine if a regional water table was at or near its normal level.

405.2.2 Artificial drainage. Areas to be monitored shall be checked for drainage tile and open ditches that alter natural high ground water levels. Where such factors are involved, information on the location, design, ownership and maintenance responsibilities for such drainage shall be provided. Documentation shall be provided to show that the drainage network has an adequate outlet and will be maintained. Sites affected by agricultural drain tile shall not be acceptable for system installation.

405.2.3 Procedures. The owner or the owner's agent shall notify the code official in writing of the intent to monitor. Where necessary, the code official shall field check the monitoring once or more during the time of expected saturated soil conditions.

Not less than three wells shall be monitored at a site for a proposed system and replacement. Where necessary, the code official shall require more than three monitoring sites, and the site evaluator shall be so advised in writing.

405.2.4 Monitoring well design. Not less than two wells shall extend to a depth of not less than 6 feet (1829 mm) below the ground surface and shall be not less than 3 feet (914 mm) below the designed system depth. However, with layered mottled soil over permeable unmottled soil, not less than one well shall terminate within the mottled layer. Monitoring at greater depths shall be required, where necessary, due to site conditions. The site evaluator shall determine the depth of the monitoring wells for each specific site. Depths shall be approved. The monitoring well shall be a solid pipe installed in a bore hole. The pipe size shall be not less than 1 inch (25 mm) and not greater than 4 inches (102 mm). The bore hole shall be not less than 4 inches (102 mm) and not greater than 8 inches (203 mm) larger than the pipe (see Appendix A, Figure A-2).

405.2.5 Observations. The first observation shall be made on or before **[DATE]**. Observations shall be made thereafter every seven days or less until **[DATE]** or until the site is determined to be unacceptable, whichever occurs first. Where water is observed above the critical depth at any time, an observation shall be made one week later. Where water is present above the critical depth at both observations, monitoring shall cease and the site shall be considered unacceptable. Where water is not present above the critical depth at the second observation, monitoring shall continue until **[DATE]**. Where any two observations seven days apart show the presence of water above the critical depth, the site shall be considered unacceptable and the code official shall be notified in writing. When rainfall of 0.5 inch (12.7 mm) or more occurs in a 24-hour period during monitoring, observations shall be made at more frequent intervals, where necessary.

405.2.6 Reporting data. Where monitoring shows saturated conditions, the following data shall be submitted in writing: test locations; ground elevations at the wells; soil profile descriptions; soil series, if available from soil maps; dates observed; depths to observed water; and local precipitation data—monthly from **[DATE]** and daily during monitoring.

Where monitoring discloses that the site is acceptable, the following data shall be submitted in writing: location and depth of test holes, ground elevations at the wells and soil profile descriptions; soil series, if available from soil maps; dates observed; results of observations; information on artificial drainage; and local precipitation data—monthly from **[DATE]** and daily during monitoring. A request to install a soil absorption system shall be made in accordance with Section 106.

SECTION 406 SITE REQUIREMENTS

406.1 Soil absorption site location. The surface grade of all soil absorption systems shall be located at a point lower than the surface grade of any nearby water well or reservoir on the same or adjoining property. Where this is not possible, the site shall be located so surface water drainage from the site is not directed toward a well or reservoir. The soil absorption system shall be located with a minimum horizontal distance between various elements as indicated in Table 406.1. Private sewage disposal systems in compacted areas, such as parking lots and driveways, are prohibited. Surface water shall be diverted away from any soil absorption site on the same or neighboring lots.

TABLE 406.1
MINIMUM HORIZONTAL SEPARATION DISTANCES FOR SOIL ABSORPTION SYSTEMS

ELEMENT	DISTANCE (feet)
Cistern	50
Habitable building, below-grade foundation	25
Habitable building, slab-on-grade	15
Lake, high-water mark	50
Lot line	5
Reservoir	50
Roadway ditches	10
Spring	100
Streams or watercourse	50
Swimming pool	15
Uninhabited building	10
Water main	50
Water service	10
Water well	50

For SI: 1 foot = 304.8 mm.

406.1.1 Flood hazard areas. The site shall be located outside of *flood hazard areas.*

Exception: Where suitable sites outside of the *flood hazard area* are not available, it is permitted for the site to be located within the *flood hazard area*. The site shall be located to minimize the effects of inundation under conditions of the design flood.

406.2 Ground water, bedrock or slowly permeable soils. There shall be not less than 3 feet (914 mm) of soil between the bottom of the soil absorption system and high ground water or bedrock. Soil with a percolation rate of 60 minutes per 1 inch (25 mm) or faster shall exist for the depth of the proposed soil absorption system and not less than 3 feet (914 mm) below the proposed bottom of the soil absorption system. There shall be 56 inches (1422 mm) of suitable soil from original grade for a *conventional soil absorption system.*

406.3 Percolation rate, trench or bed. A subsurface soil absorption system of the trench or bed type shall not be installed where the percolation rate for any one of the three tests is slower than 60 minutes for water to fall 1 inch (25 mm). The slowest percolation rate shall be used to determine the absorption area.

406.4 Percolation rate, seepage pit. Percolation tests shall be made in each horizon penetrated below the inlet pipe for a *seepage pit.* Soil strata in which the percolation rates are slower than 30 minutes per 1 inch (25 mm) shall not be included in computing the absorption area. The slowest percolation rate shall be used to determine the absorption area.

406.5 Soil maps. Where a parcel of land consists entirely of soils with very severe or severe limitations for on-site liquid-waste disposal as determined by use of a detailed soil map and supporting data, that map and supporting data shall be permitted to be used as a basis for denial for an on-site waste disposal system. However, the property owner shall be permitted to present evidence that a suitable site for an on-site liquid-waste disposal system does exist.

406.6 Filled area. A soil absorption system shall not be installed in a filled area unless written approval is received.

406.6.1 Placement of fill. The approval of a *conventional soil absorption system* shall be based on evidence indicating its conformance to code requirements for area, percolation and elevation.

406.6.2 Bedrock. Where the original soil texture is sand or loamy sand, and the site has not less than 30 inches (762 mm) and not greater than 56 inches (1422 mm) of soil over bedrock, the fill shall be the same or coarser soil texture as the natural soil. Coarser fill material shall not be coarser than medium sand. Fill material shall not be finer than the natural soil.

406.6.3 High ground water. Sites with less than 56 inches (1422 mm) of soil over high ground water or estimated high ground water, where the original soil texture is sand or loamy sand, are permitted to be filled in accordance with Section 406.6.1 or 406.6.2.

406.6.4 Natural soil. Sites with soils finer than sand or loamy sand shall not be approved for systems in fill.

406.6.5 Monitoring. Sites that will have 36 inches (762 mm) or less of soil above high ground water after the topsoil is removed shall be monitored for high ground water levels in the filled area in accordance with Section 405.2.

406.6.6 Inspection of fill. Placement of the fill material shall be inspected by the code official.

406.6.7 Design requirements. Filled areas shall be large enough to accommodate a shallow trench system and a replacement system. The site of the area to be filled shall be determined by the percolation rate of the natural soil and use of the building. Where any portion of the trench system or its replacement is in the fill, the fill shall extend 20 feet (6096 mm) beyond all sides of both systems before the slope begins. *Soil borings* and percolation tests shall be conducted before filling to determine soil textures and depth to high ground water or bedrock. Vegetation and topsoil shall be removed prior to filling. Slopes at the edge of the filled areas shall have a maximum ratio of one unit vertical to three units horizontal (33-percent slope), provided that the 20-foot (6096 mm) separating distance is maintained (see Appendix A, Figure A-3).

406.7 Altering slopes. Areas with slopes exceeding those specified in Section 402.1 shall not be used unless graded and reshaped in accordance with Sections 406.7.1 through 406.7.3.

406.7.1 Site investigation. Soil test data shall show that a sufficient depth of suitable soil material is present to provide the required amount of soil over bedrock and ground water after alteration. A complete site evaluation as specified in this section shall be performed after alteration of the site.

406.7.2 System location. A soil absorption system shall be installed in the cut area of an altered site. A soil absorption system shall not be installed in the fill area of an altered site. The area of fill on an altered site is permitted to be used as a portion of the required 20-foot (6096 mm) separating distance from the crown of a critical slope. There shall be not less than 6 feet (1829 mm) of natural soil between the edge of a system area and the downslope side of the altered area.

406.7.3 Site protection. Altered slope areas shall be positioned so that surface water drainage will be diverted away from the system areas. Disturbed areas shall be seeded or sodded with grass, and appropriate steps shall be taken to control erosion (see Figure 406.7.3).

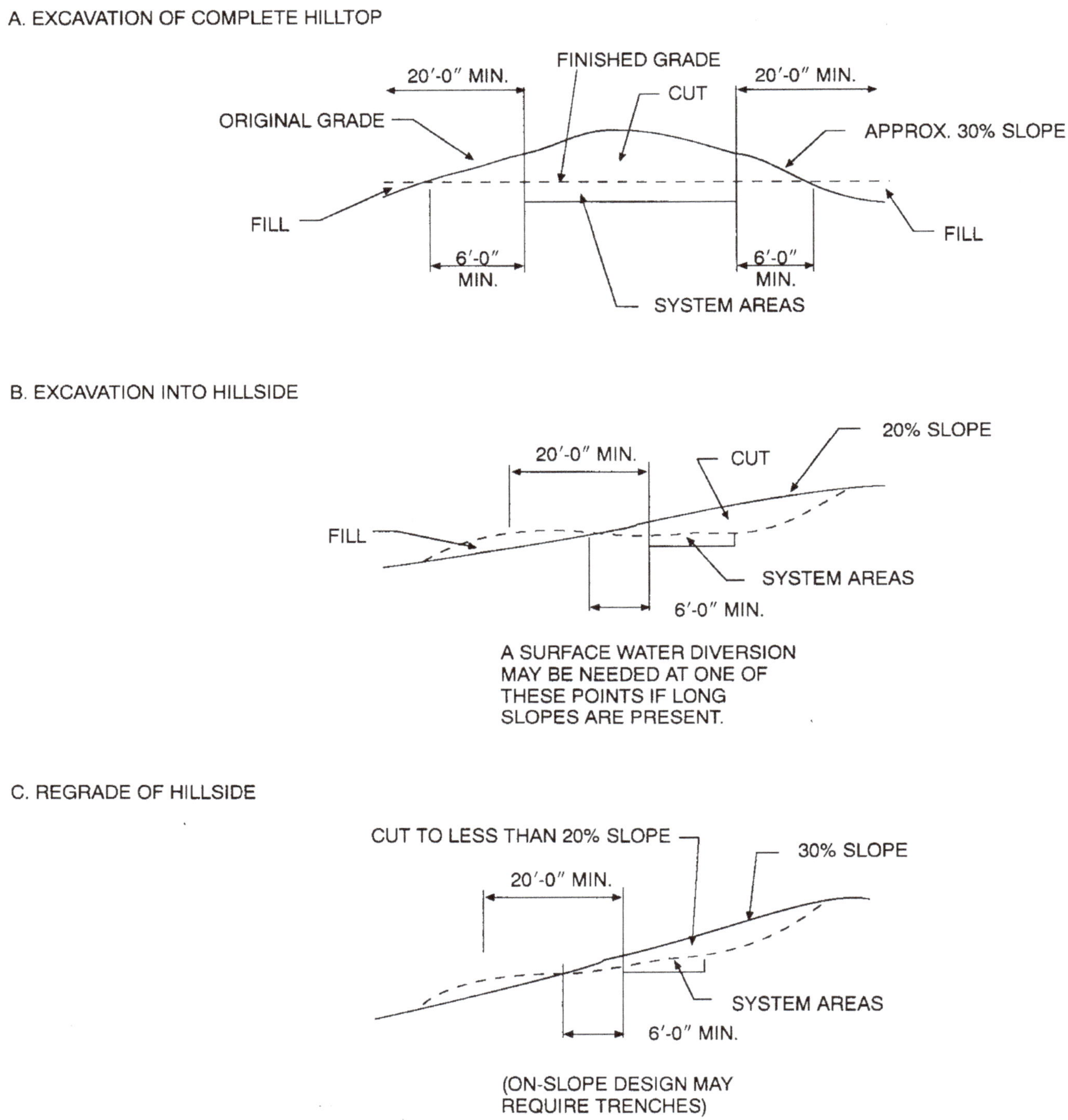

For SI: 1 inch = 25.4 mm, 1 foot = 304.8 mm.

FIGURE 406.7.3
CONCEPTUAL DESIGN SKETCH FOR ALTERING SLOPES

CHAPTER 5

MATERIALS

User note:

About this chapter: *Piping materials used in private sewage disposal systems must comply with standards. Chapter 5 indicates the standards for these products and specifies the material requirements for steel, concrete and fiberglass tanks.*

SECTION 501 GENERAL

501.1 Scope. The provisions of this chapter shall govern the requirements for materials for *private sewage disposal systems*.

501.2 Minimum standards. Materials shall conform to the standards referenced in this code for the construction, installation, alteration or repair of *private sewage disposal systems* or parts thereof.

> **Exception:** The extension, addition to or relocation of existing pipes with materials of like grade or quality in accordance with Sections 102.6 and 105.

SECTION 502 IDENTIFICATION

502.1 General. The manufacturer's mark or name and the quality of the product or identification shall be cast, embossed, stamped or indelibly marked on each length of pipe and each pipe fitting, fixture, tank, material and device used in a *private sewage disposal system* in accordance with the approved standard. Tanks shall indicate their capacity.

SECTION 503 PERFORMANCE REQUIREMENTS

503.1 Approved materials required. Materials, fixtures or equipment used in the installation, repair or alteration of any *private sewage disposal system* shall conform to the standards referenced in this code, except as otherwise approved in accordance with Section 105.

503.2 Care in installation. Materials installed in *private sewage disposal systems* shall be handled and installed so as to avoid damage. The quality of the material shall not be impaired.

503.3 Defective materials prohibited. Defective or damaged materials, equipment or apparatus shall not be installed or maintained.

SECTION 504 TANKS

504.1 Approval. Tanks shall be of an approved type. The design of tanks shall conform to the requirements of Chapter 8. Tanks shall be designed to withstand the pressures to which they are subjected.

504.2 Precast concrete and site-constructed tanks. Precast concrete tanks shall conform to ASTM C913. The floor and sidewalls of a site-constructed concrete tank shall be monolithic, except a construction joint is permitted in the lower 12 inches (305 mm) of the sidewalls of the tank. The construction joint shall have a keyway in the lower section of the joint. The width of the keyway shall be approximately 30 percent of the thickness of the sidewall with a depth equal to the width. A continuous water stop or baffle not less than 56 inches (1422 mm) wide shall be set vertically in the joint, embedded one-half its width in the concrete below the joint with the remaining width in the concrete above the joint. The water stop or baffle shall be copper, neoprene, rubber or polyvinyl chloride designed for this specific purpose. Joints between the concrete septic tank and the tank cover and between the septic tank cover and manhole riser shall be tongue and groove or shiplap-type and sealed water tight using cement, mortar or bituminous compound.

504.3 Steel tanks. Steel tanks shall conform to UL 70. Any damage to the bituminous coating shall be repaired by recoating. The gage of the steel shall be in accordance with Table 504.3.

TABLE 504.3 TANK CAPACITY

TANK DESIGN AND CAPACITY		MINIMUM GAGE THICKNESS	MINIMUM DIAMETER
Vertical cylindrical			
500 to 1,000 gallons	Bottom and sidewalls	12 gage	None
	Cover	12 gage	
	Baffles	12 gage	
1,001 to 1,250 gallons	Complete tank	10 gage	None
1,251 to 1,500 gallons	Complete tank	7 gage	None
Horizontal cylindrical			
500 to 1,000 gallons	Complete tank	12 gage	54-inch diameter
1,001 to 1,500 gallons	Complete tank	12 gage	64-inch diameter
1,501 to 2,500 gallons	Complete tank	10 gage	76-inch diameter
2,501 to 9,000 gallons	Complete tank	7 gage	76-inch diameter
9,001 to 12,000 gallons	Complete tank	$^{1}/_{4}$-inch plate	None
Over 12,000 gallons	Complete tank	$^{5}/_{16}$ inch	None

For SI: 1 inch = 25.4 mm, 1 gallon = 3.785 L.

504.4 Fiberglass tanks. Fiberglass tanks shall conform to ASTM D4021.

504.5 Manholes. Manhole collars and extensions shall be of the same material as the tank. Manhole covers shall be of concrete, steel, cast iron or other approved material.

SECTION 505
PIPE, JOINTS AND CONNECTIONS

505.1 Pipe. Pipe for *private sewage disposal systems* shall have a smooth wall and conform to one of the standards listed in Table 505.1.

505.1.1 Distribution pipe. Perforated pipe for distribution systems shall conform to one of the standards listed in Table 505.1 or 505.1.1.

505.2 Joints and connection approval. Joints and connections shall be of an approved type.

505.3 ABS plastic pipe. Joints between acrylonitrile butadiene styrene (ABS) plastic pipe or fittings shall be in accordance with Sections 505.3.1 and 505.3.2.

505.3.1 Mechanical joints. Mechanical joints on drainage pipes shall be made with an elastomeric seal conforming to ASTM C1173, ASTM D3212 or CSA B 602. Mechanical joints shall be installed only in underground systems, except as otherwise approved. Joints shall be installed in accordance with the manufacturer's instructions.

505.3.2 Solvent cementing. Joint surfaces shall be clean and free from moisture. Solvent cement conforming to ASTM D2235 or CSA B181.1 shall be applied to all joint surfaces. The joint shall be made while the cement is wet. Joints shall be made in accordance with ASTM D2235, ASTM D2661, ASTM F628 or CSA B181.1. Solvent cement joints shall be permitted above or below ground.

505.4 Asbestos-cement pipe. Joints between asbestos-cement pipe or fittings shall be made with a sleeve coupling of the same composition as the pipe and sealed with an elastomeric ring conforming to ASTM D1869.

505.5 Coextruded composite ABS pipe and joints. Joints between coextruded composite pipe with an ABS outer layer or ABS fittings shall comply with Sections 505.5.1 and 505.5.2.

505.5.1 Mechanical joints. Mechanical joints on drainage pipe shall be made with an elastomeric seal conforming to ASTM C1173, ASTM D3212 or CSA B 602. Mechanical joints shall not be installed in above-ground systems, except as otherwise approved. Joints shall be installed in accordance with the manufacturer's instructions.

505.5.2 Solvent cementing. Joint surfaces shall be clean and free from moisture. Solvent cement conforming to ASTM D2235 or CSA B 181.1 shall be applied to all joint surfaces. The joint shall be made while the cement is wet. Joints shall be made in accordance with ASTM D2235, ASTM D2661, ASTM F628 or CSA B181.1. Solvent cement joints shall be permitted above or below ground.

505.6 Cast-iron pipe. Joints between cast-iron pipe or fittings shall be in accordance with Sections 505.6.1 through 505.6.3.

TABLE 505.1
PRIVATE SEWAGE DISPOSAL SYSTEM PIPE

MATERIAL	STANDARD
Acrylonitrile butadiene styrene (ABS) plastic pipe	ASTM D2661; ASTM D2751; ASTM F628
Asbestos-cement pipe	ASTM C428
Cast-iron pipe	ASTM A74; ASTM A888; CISPI 301
Coextruded composite ABS DWV Schedule 40 IPS pipe (solid)	ASTM F1488; ASTM F1499
Coextruded composite ABS DWV Schedule 40 IPS pipe (cellular core)	ASTM F1488; ASTM F1499
Coextruded composite ABS sewer and drain DR-PS in PS35, PS50, PS100, PS140 and PS200	ASTM F1488; ASTM F1499
Coextruded composite PVC DWV Schedule 40 IPS pipe (solid)	ASTM F1488
Coextruded composite PVC DWV Schedule 40 IPS pipe (cellular core)	ASTM F1488
Coextruded composite PVC-IPS-DR of PS140, PS200, DWV	ASTM F1488
Coextruded composite PVC 3.25 OD DWV pipe	ASTM F1488
Coextruded composite PVC sewer and drain DR-PS in PS35, PS50, PS100, PS140 and PS200	ASTM F1488
Concrete pipe	ASTM C14; ASTM C76; CSA A257.1M; CSA A257.2M
Copper or copper-alloy tubing (Type K or L)	ASTM B75; ASTM B88; ASTM B251
Polyvinyl chloride (PVC) plastic pipe (Type DWV, SDR26, SDR35, SDR41, PS50 or PS100)	ASTM D2665; ASTM D2949; ASTM D3034; ASTM F891; CSA B182.2; CSA B182.4
Vitrified clay pipe	ASTM C4; ASTM C700

TABLE 505.1.1
DISTRIBUTION PIPE

MATERIAL	STANDARD
Polyethylene (PE) plastic pipe	ASTM F405
Polyvinyl chloride (PVC) plastic pipe	ASTM D2729
Polyvinyl chloride (PVC) plastic pipe with pipe stiffness of PS35 and PS50	ASTM F1488

505.6.1 Caulked joints. Joints for hub and spigot pipe shall be firmly packed with oakum or hemp. Molten lead shall be poured in one operation to a depth of not less than 1 inch (25 mm). The lead shall not recede more than 0.125 inch (3.2 mm) below the rim of the hub, and shall be caulked tight. Paint, varnish or other coatings shall not be applied to the joining material until after the joint has been tested and approved. Lead shall be run in one pouring and shall be caulked tight. Acid-resistant rope and acidproof cement shall be permitted.

505.6.2 Mechanical compression joints. Compression gaskets for hub and spigot pipe and fittings shall conform to ASTM C564. Gaskets shall be compressed when the pipe is fully inserted.

505.6.3 Mechanical joint coupling. Mechanical joint couplings for hubless pipe and fittings shall comply with CISPI 310 or ASTM C1277. The elastomeric sealing sleeve shall conform to ASTM C564 or CSA B 602 and shall be provided with a center stop. Mechanical joint couplings shall be installed in accordance with the manufacturer's instructions.

505.7 Concrete pipe. Joints between concrete pipe or fittings shall be made by the use of an elastomeric seal conforming to ASTM C443, ASTM C1173, CSA A 257.3M or CSA B 602.

505.8 Copper or copper-alloy tubing or pipe. Joints between copper or copper-alloy tubing, pipe or fittings shall be in accordance with Sections 505.8.1 and 505.8.2.

505.8.1 Mechanical joints. Mechanical joints shall be installed in accordance with the manufacturer's instructions.

505.8.2 Soldered joints. Solder joints shall be made in accordance with the methods of ASTM B828. Cut ends shall be reamed to the full inside diameter of the tube end. All joint surfaces shall be cleaned. A flux conforming to ASTM B813 shall be applied. The joint shall be soldered with a solder conforming to ASTM B32.

505.9 Polyethylene plastic pipe and tubing. Joints between polyethylene plastic pipe and tubing or fittings shall be in accordance with Sections 505.9.1 and 505.9.2.

505.9.1 Heat-fusion joints. Joint surfaces shall be clean and free from moisture. Joint surfaces shall be heated to melting temperature and joined. The joint shall be undisturbed until cool. Joints shall be made in accordance with ASTM D2657.

505.9.2 Mechanical joints. Mechanical joints shall be installed in accordance with the manufacturer's instructions.

505.10 PVC plastic pipe. Joints between polyvinyl chloride (PVC) plastic pipe and fittings shall be in accordance with Sections 505.10.1 and 505.10.2.

505.10.1 Mechanical joints. Mechanical joints shall be made with an elastomeric seal conforming to ASTM C1173, ASTM D3212 or CSA B602. Mechanical joints shall not be installed in above-ground systems, except as otherwise approved. Joints shall be installed in accordance with the manufacturer's instructions.

505.10.2 Solvent cementing. Joint surfaces shall be clean and free from moisture. A purple primer that conforms to ASTM F656 shall be applied. Solvent cement not purple in color and conforming to ASTM D2564, CSA B137.3, CSA B181.2 or CSA B182.1 shall be applied to all joint surfaces. The joint shall be made while the cement is wet, and shall be in accordance with ASTM D2855. Solvent cement joints shall be permitted above or below ground.

505.11 Coextruded composite PVC pipe. Joints between coextruded composite pipe with a PVC outer layer or PVC fittings shall comply with Sections 505.11.1 and 505.11.2.

505.11.1 Mechanical joints. Mechanical joints on drainage pipe shall be made with an elastomeric seal conforming to ASTM D3212. Mechanical joints shall not be installed in above-ground systems, except as otherwise approved. Joints shall be installed in accordance with the manufacturer's instructions.

505.11.2 Solvent cementing. Joint surfaces shall be clean and free from moisture. A purple primer that conforms to ASTM F656 shall be applied. Solvent cement not purple in color and conforming to ASTM D2564, CSA B137.3, CSA B181.2 or CSA B 182.1 shall be applied to all joint surfaces. The joint shall be made while the cement is wet, and shall be in accordance with ASTM D2855. Solvent cement joints shall be permitted above or below ground.

505.12 Vitrified clay pipe. Joints between vitrified clay pipe or fittings shall be made by the use of an elastomeric seal conforming to ASTM C425, ASTM C1173 or CSA B602.

505.13 Different piping materials. Joints between different piping materials shall be made with a mechanical joint of the compression or mechanical-sealing type conforming to ASTM C1173, ASTM C1460 or ASTM C1461. Connectors or adapters shall be approved for the application and such joints shall have an elastomeric seal conforming to ASTM C425, ASTM C443, ASTM C564, ASTM C1440, ASTM D1869, ASTM F477, CSA A257.3M or CSA B602 or as required in Sections 505.13.1 and 505.13.2. Joints shall be installed in accordance with the manufacturer's instructions.

505.13.1 Copper to cast-iron hub pipe. Joints between copper pipe or copper alloy tubing and cast-iron hub pipe shall be made with a copper-alloy ferrule or compression joint. The copper pipe or tubing shall be soldered to the ferrule in an approved manner, and the ferrule shall be joined to the cast-iron hub by a caulked joint or a mechanical compression joint.

505.13.2 Plastic pipe or tubing to other piping material. Joints between different grades of plastic pipe or between plastic pipe and other piping material shall be made with an approved adapter fitting. Joints between plastic pipe and cast-iron hub pipe shall be made by a caulked joint or a mechanical compression joint.

505.14 Pipe installation. Pipe shall be installed in accordance with the *International Plumbing Code.*

SECTION 506
PROHIBITED JOINTS AND CONNECTIONS

506.1 General. The following types of joints and connections shall be prohibited:

1. Cement or concrete joints.
2. Mastic or hot-pour bituminous joints.

3. Joints made with fittings not approved for the specific installation.
4. Joints between different diameter pipes made with elastomeric rolling O-rings.
5. Solvent-cement joints between different types of plastic pipe.

CHAPTER 6

SOIL ABSORPTION SYSTEMS

User note:

About this chapter: *Chapter 6 covers the design of 'conventional' soil absorption systems. These systems are conventional in the sense that nonpotable water from the outlet of a septic tank flows by gravity into a piping network for distributing the water in an excavated area nearby. The piping is backfilled with the finished grade blending into adjacent grade level.*

SECTION 601 GENERAL

601.1 Scope. The provisions of this chapter shall govern the sizing and installation of soil absorption systems.

SECTION 602 SIZING SOIL ABSORPTION SYSTEMS

602.1 General. Effluent from septic tanks and other approved treatment tanks shall be disposed of by soil absorption or an approved manner. Sizing shall be in accordance with this chapter for systems with a daily effluent application of 5,000 gallons (18 925 L) or less. Two systems of equal size shall be required for systems receiving effluents exceeding 5,000 gallons (18 925 L) per day. Each system shall have a minimum capacity of 75 percent of the area required for a single system. An approved means of alternating waste application shall be provided. A dual system shall be considered as one system.

602.2 Pressure system. A *pressure distribution system* shall be permitted in place of a conventional or dosing *conventional soil absorption system* where a site is suitable for a conventional *private sewage disposal system*. A *pressure distribution system* shall be approved as an alternative *private sewage disposal system* where the site is unsuitable for conventional treatment (for sizing and design criteria, see Chapter 7).

602.3 Method of discharge. Flow from the septic or treatment tank to the soil absorption system shall be by gravity or dosing for facilities with a daily effluent application of 1,500 gallons (5678 L) or less. The tank effluent shall be discharged by pumping or an automatic siphon for systems over 1,500 gallons (5678 L).

SECTION 603 RESIDENTIAL SIZING

603.1 General. The bottom area for *seepage trenches* or beds or the sidewall area for *seepage pits* required for a soil absorption system serving residential property shall be determined from Table 603.1 using soil percolation test data and type of construction.

TABLE 603.1 MINIMUM ABSORPTION AREA FOR ONE- AND TWO-FAMILY DWELLINGS

PERCOLATION CLASS	PERCOLATION RATE (minutes required for water to fall 1 inch)	SEEPAGE TRENCHES OR PITS (square feet per bedroom)	SEEPAGE BEDS (square feet per bedroom)
1	0 to less than 10	165	205
2	10 to less than 30	250	315
3	30 to less than 45	300	375
4	45 to 60	330	415

For SI: 1 minute per inch = 2.4 s/mm, 1 square foot = 0.0929 m^2.

SECTION 604 OTHER BUILDING SIZING

604.1 General. The minimum required soil absorption system area for all occupancies, except one- and two-family dwellings, shall be based on building usage, the percolation rate and system design in accordance with Tables 604.1(1) and 604.1(2). The minimum soil absorption area shall be calculated by the following equation:

$$A = U \times CF \times AA \qquad \textbf{(Equation 6-1)}$$

where:

A = Minimum system absorption area.

AA = Absorption area from Table 604.1(1).

CF = Conversion factor from Table 604.1(2).

U = Number of units.

TABLE 604.1(1) MINIMUM ABSORPTION AREA FOR OTHER THAN ONE- AND TWO-FAMILY DWELLINGS

PERCOLATION CLASS	PERCOLATION RATE (minutes required for water to fall 1 inch)	SEEPAGE TRENCHES OR PITS (square feet per unit)	SEEPAGE BEDS (square feet per unit)
1	0 to less than 10	110	140
2	10 to less than 30	165	205
3	30 to less than 45	220	250
4	45 to 60	220	280

For SI: 1 minute per inch = 2.4 s/mm, 1 square foot = 0.0929 m^2.

TABLE 604.1(2)
CONVERSION FACTOR

BUILDING CLASSIFICATION	UNITS	FACTOR
Apartment building	1 per bedroom	1.5
Assembly hall—no kitchen	1 per person	0.02
Auto washer (service buildings, etc.)	1 per machine	6.0
Bar and cocktail lounge	1 per patron space	0.2
Beauty salon	1 per station	2.4
Bowling center	1 per bowling lane	2.5
Bowling center with bar	1 per bowling lane	4.5
Camp, day and night	1 per person	0.45
Camp, day use only	1 per person	0.2
Campground and camping resort	1 per camping space	0.9
Campground and sanitary dump station	1 per camping space	0.085
Car wash	1 per car	1.0
Catch basin—garages, motor-fuel-dispensing facility, etc.	1 per basin	2.0
Catch basin—truck wash	1 per truck	5.0
Church—no kitchen	1 per person	0.04
Church—with kitchen	1 per person	0.09
Condominium	1 per bedroom	1.5
Dance hall	1 per person	0.06
Dining hall—kitchen and toilet	1 per meal served	0.2
Dining hall—kitchen and toilet waste with dishwasher or food waste grinder or both	1 per meal served	0.25
Dining hall—kitchen only	1 per meal served	0.06
Drive-in restaurant, inside seating	1 per seat	0.3
Drive-in restaurant, without inside seating	1 per car space	0.3
Drive-in theater	1 per car space	0.1
Employees—in all buildings	1 per person	0.4
Floor drain	1 per drain	1.0
Hospital	1 per bed space	2.0
Hotel or motel and tourist rooming house	1 per room	0.9
Labor camp—central bathhouse	1 per employee	0.25
Medical office buildings, clinics and dental offices Doctors, nurses and medical staff Office personnel Patients	 1 per person 1 per person 1 per person	 0.8 0.25 0.15
Mobile home park	1 per mobile home site	3.0
Motor-fuel-dispensing facility	1 per car served	0.15
Nursing or group homes	1 per bed space	1.0
Outdoor sports facility—toilet waste only	1 per person	0.35
Park—showers and toilets	1 per acre	8.0
Park—toilet waste only	1 per acre	4.0
Restaurant—dishwasher or food waste grinder or both	1 per seating space	0.15
Restaurant—kitchen and toilet	1 per seating space	0.6

(continued)

TABLE 604.1(2)—continued
CONVERSION FACTOR

BUILDING CLASSIFICATION	UNITS	FACTOR
Restaurant—kitchen waste only	1 per seating space	0.18
Restaurant—toilet waste only	1 per seating space	0.42
Restaurant—(24-hour) kitchen and toilet	1 per seating space	1.2
Restaurant—(24-hour) with dishwasher or food waste grinder or both	1 per seating space	1.5
Retail store	1 per customer	0.03
School—meals and showers	1 per classroom	8.0
School—meals served or showers	1 per classroom	6.7
School—no meals, no showers	1 per classroom	5.0
Self-service laundry—toilet waste only	1 per machine	1.0
Showers—public	1 per shower	0.3
Swimming pool bathhouse	1 per person	0.2

SECTION 605
INSTALLATION OF CONVENTIONAL SOIL ABSORPTION SYSTEMS

605.1 Seepage trench excavations. *Seepage trench* excavations shall be 1 foot to 5 feet (305 mm to 1524 mm) wide. Trench excavations shall be spaced not less than 6 feet (1829 mm) apart. The absorption area of a *seepage trench* shall be computed by using only the bottom of the trench area. The bottom excavation area of the distribution header shall not be computed as absorption area. Individual *seepage trenches* shall be not greater than 100 feet (30 480 mm) long, except as otherwise approved.

605.2 Seepage bed excavations. *Seepage bed* excavations shall be not less than 5 feet (1524 mm) wide and have more than one distribution pipe. The absorption area of a *seepage bed* shall be computed by using the bottom of the trench area. Distribution piping in a *seepage bed* shall be uniformly spaced not greater than 5 feet (1524 mm) and not less than 3 feet (914 mm) apart, and not greater than 3 feet (914 mm) and not less than 1 foot (305 mm) from the sidewall or headwall.

605.3 Seepage pits. A *seepage pit* shall have not less than an inside diameter of 5 feet (1524 mm) and shall consist of a chamber walled-up with material, such as perforated precast concrete ring, concrete block, brick or other approved material allowing effluent to percolate into the surrounding soil. The pit bottom shall be left open to the soil. Aggregate of $^1/_2$ inch to $2^1/_2$ inches (12.7 mm to 64 mm) in size shall be placed into a 6-inch minimum (152 mm) annular space separating the outside wall of the chamber and sidewall excavation. The depth of the annular space shall be measured from the inlet pipe to the bottom of the chamber. Each *seepage pit* shall be provided with a 24-inch (610 mm) manhole extending to within 56 inches (1422 mm) of the ground surface and a 4-inch-diameter (102 mm) fresh air inlet. *Seepage pits* shall be located not less than 5 feet (1524 mm) apart. Excavation and scarifying shall be in accordance with Section 605.4. The effective area of a *seepage pit* shall be the vertical wall area of the walled-up chamber for the depth below the inlet for all strata in which the percolation rates are less than 30 minutes per inch (70 s/mm). The 6-inch (152 mm) annular opening outside the vertical wall area is permitted to be included for determining the effective area. Table 605.3, or an approved method, shall be used for determining the effective sidewall area of circular *seepage pits*.

TABLE 605.3
EFFECTIVE SQUARE-FOOT ABSORPTION AREA FOR SEEPAGE PITS

INSIDE DIAMETER OF CHAMBER IN FEET PLUS 1 FOOT FOR WALL THICKNESS PLUS 1 FOOT FOR ANNULAR SPACE	DEPTH IN FEET OF PERMEABLE STRATA BELOW INLET					
	3	4	5	6	7	8
7	47	88	110	132	154	176
8	75	101	126	151	176	201
9	85	113	142	170	198	226
10	94	126	157	188	220	251
11	104	138	173	208	242	277
13	123	163	204	245	286	327

For SI: 1 foot = 304.8 mm.

605.4 Excavation and construction. The bottom of a trench or bed excavation shall be level. *Seepage trenches* or beds shall not be excavated where the soil is so wet that such material rolled between the hands forms a soil wire. Smeared or compacted soil surfaces in the sidewalls or bottom of *seepage trench* or bed excavations shall be scarified to the depth of smearing or compaction and the loose material removed. Where rain falls on an open excavation, the soil shall be left until sufficiently dry so a soil wire will not form when soil from the excavation bottom is rolled between the hands. The bottom area shall then be scarified and loose material removed.

605.5 Aggregate and backfill. Not less than 6 inches (152 mm) of aggregate ranging in size from $^1/_2$ inch to $2^1/_2$ inches (12.7 mm to 64 mm) shall be laid into the trench or bed below the distribution pipe elevation. The aggregate shall be evenly distributed not less than 2 inches (51 mm) over the top of the distribution pipe. The aggregate shall be covered with approved synthetic materials or 9 inches (229 mm) of uncompacted marsh hay or straw. Building paper shall not be used to cover the aggregate. Not less than 18 inches (457 mm) of soil backfill shall be provided above the covering.

605.6 Distribution piping. Distribution piping for gravity systems shall be not less than 4 inches (102 mm) in diameter. The distribution header (PVC) shall be solid-wall pipe. The top of the distribution pipe shall be not less than 8 inches (203 mm) below the original surface in continuous straight or curved lines. The slope of the distribution pipes shall be 2 inches to 4 inches (51 mm to 102 mm) per 100 feet (30 480 mm). Effluent shall be distributed to all distribution pipes. Distribution of effluent to *seepage trenches* on sloping sites shall be accomplished by using a drop box design or other approved methods. Where dosing is required, the siphon or pump shall discharge a dose of minimum capacity equal to 75 percent of the combined volume of the distribution piping in the absorption system.

605.7 Observation pipes. Observation pipes shall be provided. Such pipes shall be not less than 4 inches (102 mm) in diameter, not less than 12 inches (305 mm) above final grade and shall terminate with an approved vent cap.

The bottom 12 inches (305 mm) of the observation pipe shall be perforated and extend to the bottom of the aggregate. Observation pipes shall be located not less than 25 feet (7620 mm) from any window, door or air intake of any building used for human occupancy. Not more than four distribution pipelines shall be served by one common 4-inch (102 mm) observation pipe where interconnected by a common header pipe (see Appendix-A, Figure A-4).

Exception: Where approved and where the location of the observation pipe is permanently recorded, the observation pipe shall be not more than 2 inches (51 mm) below the finished grade.

605.8 Winter installation. Soil absorption systems shall not be installed during periods of adverse weather conditions unless the installation is approved. A soil absorption system shall not be installed where the soil at the system elevation is frozen. Snow cover shall be removed from the soil absorption area before excavation begins. Snow shall not be placed in a manner that will cause water to pond on the soil absorption system area during snow melt. Excavated soil to be used as backfill shall be protected from freezing. Excavated soil that freezes solid shall not be used as backfill. The first 12 inches (305 mm) of backfill shall be loose, unfrozen soil. Inspection of systems installed during winter conditions shall include inspection of the trench or bed excavation prior to the placement of gravel and inspection of backfill material at the time of placement.

605.9 Evaporation. Soil absorption systems shall not be covered or paved over by material that inhibits the evaporation of the effluent.

CHAPTER 7
PRESSURE DISTRIBUTION SYSTEMS

User note:

About this chapter: *Chapter 7 regulates the pressure distribution method of soil absorption systems. This type of system is necessary where the water from the outlet of a septic tank cannot flow by gravity because of site constraints. The water is pumped from a collection tank to the absorption area at regular intervals.*

SECTION 701
GENERAL

701.1 Scope. The provisions of this chapter shall govern the design and installation of *pressure distribution systems*.

SECTION 702
DESIGN LOADING RATE

702.1 General. A *pressure distribution system* shall be permitted for use on any site meeting the conventional *private sewage disposal system* criteria. There shall be not less than 6 inches (152 mm) to the top of the distribution piping from original grade for any *pressure distribution system*. The minimum required suitable soil depths from original grade for *pressure distribution systems* shall be in accordance with Table 702.1.

TABLE 702.1
SOIL REQUIRED

DISTRIBUTION PIPE (inches)	SUITABLE SOIL (inches)
1	49
2	50
3	52
4	53

For SI: 1 inch = 25.4 mm.

702.2 Absorption area. The total absorption area required shall be computed from the estimated daily wastewater flow and the design loading rate based on the percolation rate for the site. The required absorption area equals wastewater flow divided by the design loading rate from Table 702.2. Two systems of equal size shall be required for systems receiving effluents exceeding 5,000 gallons (18 925 L). Each system shall have a minimum capacity of 75 percent of the area required for a single system and shall be provided with a suitable means of alternating waste applications. A dual system shall be considered as one system.

TABLE 702.2
DESIGN LOADING RATE

PERCOLATION RATE (minutes per inch)	DESIGN LOADING FACTOR (gallons per square foot per day)
0 to less than 10	1.2
10 to less than 30	0.8
30 to less than 45	0.72
45 to 60	0.4

For SI: 1 minute per inch = 2.4 s/mm, 1 gallon per square foot = 0.025 L/m^2.

702.3 Estimated wastewater flow. The estimated wastewater flow from a residence shall be 150 gallons (568 L) per bedroom per day. Wastewater flow rates for other occupancies in a 24-hour period shall be based on the values in Table 802.7.2.

SECTION 703
SYSTEM DESIGN

703.1 General. *Pressure distribution systems* shall discharge effluent into trenches or beds. Each pipe connected to an outlet of a manifold shall be counted as a separate distribution pipe. The horizontal spacing of distribution pipes shall be 30 inches to 72 inches (762 mm to 1829 mm). The system shall be sized in accordance with the formulas listed in this section. Systems using Schedule 40 plastic pipe shall be sized in accordance with the formulas listed in this section or in accordance with the tables listed in Appendix B. Distribution piping shall be installed at the same elevation, unless an approved system provides for a design ensuring equal flow through each of the perforations and the effluent is uniformly applied to the soil infiltrative surface (see Appendix A, Figure A-5).

703.2 Symbols. The following symbols and notations shall apply to the provisions of this chapter:

C_h = Hazen-Williams friction factor.

D = Distribution pipe diameter, inches (mm).

d = Perforation diameter, inches (mm).

D_d = Delivery pipe diameter, inches (mm).

D_m = Manifold pipe diameter, inches (mm).

f = Fraction of total head loss in the manifold segment.

F_D = Friction loss in the delivery pipe, feet of head (mm of head).

F_i = Friction factor for i^{th} manifold segment.

F_N = Friction loss in the network pipe, feet of head (mm of head).

h = Pressure in distribution pipe, feet of head (mm of head).

h_d = In-line pressure at distal end of lateral, feet of head (mm of head).

L_D = Length of delivery pipe, feet (mm).

L_i = Length of ith manifold segment, feet (mm).

N = Number of perforations in the lateral.

q = Perforation discharge rate, gpm (L/min).

Q_i = Flow rate i^{th} manifold segment, gpm (L/min).

Q_m = Flow rate at manifold inlet, gpm (L/min).

703.3 Distribution pipe. Distribution pipe size, hole diameter and hole spacing shall be selected. The hole diameter and spacing shall be equal for each manifold segment. Distribution pipe size shall not be required to be the same for each segment. Changes in pressure in the distribution pipe shall be less than or equal to 10 percent by conforming to the following formula:

$$\sum \Delta h 0.2 h_d \qquad \textbf{(Formula 7-1)}$$

For SI: 1 foot = 304.8 mm.

where:

$$\Delta h = 4.71L\left(\frac{q}{C_h D^{2.65}}\right)^{1.85}$$

$$q = 11.79d^2\sqrt{h_d}$$

The Hazen-Williams friction factor, C_h, for each pipe material shall be determined in accordance with Table 703.3.

**TABLE 703.3
HAZEN-WILLIAMS FRICTION FACTOR**

MATERIAL	FRICTION FACTOR, C_h
ABS plastic pipe	150
Asbestos-cement pipe	140
Bituminized fiber pipe	120
Cast-iron pipe	100
Concrete pipe	110
Copper or copper-alloy tubing	150
PVC plastic pipe	150
Vitrified clay pipe	100

703.4 Manifolds. The diameter of the manifold pipe shall be determined by the following equation:

$$D_m = \left(\frac{\sum L_i F_i}{f h_d}\right)^{0.21} \qquad \textbf{(Equation 7-1)}$$

For SI: 1 inch = 25.4 mm.

where:

F_i = $9.8 \times 10^{-4}\, Q_i$

q = $11.79d^2\sqrt{h_d}$

Q_i = Nq

The fraction of the total head loss at the manifold segment, f, shall be less than or equal to 0.1. The in-line pressure at the distal end of the lateral, h_d, shall be not less than 2.5 feet (762 mm) of head. Distribution pipes shall be connected to the manifold with tees or 90-degree (1.57 rad) ells. Distribution pipes shall have the ends capped.

703.5 Friction loss. The delivery pipe shall include all pipe between the pump and the supply end of the distribution pipe. The friction loss in the delivery pipe, F_D, shall be determined by the following equation:

$$F_D = L_D\left(\frac{3.55Q_m}{C_h D_d^{2.63}}\right)^{1.85} \qquad \textbf{(Equation 7-2)}$$

For SI: 1 inch of head = 25.4 mm of head.

The Hazen-Williams friction factor, C_h, for each pipe material shall be determined in accordance with Table 703.3.

The friction loss in the network pipe shall be determined by the following equation:

$$F_N = 1.31\, h_d \qquad \textbf{(Equation 7-3)}$$

For SI: 1 inch of head = 25.4 mm of head.

Pipe in the system shall be increased in size if the friction loss is excessive.

703.6 Force main. Size of the force main between the pump and manifold shall be based on the friction loss and velocity of effluent through the pipe. The velocity of effluent in a force main shall be not more than 5 feet per second (1524 mm/sec).

SECTION 704
BED AND TRENCH CONSTRUCTION

704.1 General. The excavation and construction for *pressure distribution system* trenches and beds shall be in accordance with Chapter 6. Aggregate shall be not less than 6 inches (152 mm) beneath the distribution pipe with 2 inches (51 mm) spread evenly above the pipe. The aggregate shall be clean, nondeteriorating 0.5-inch to 2.5-inch (12.7 mm to 64 mm) stone.

SECTION 705
PUMPS

705.1 General. Pump selection shall be based on the discharge rate and total dynamic head of the pump performance curve. The total dynamic head shall be equal to the difference in feet (mm) of elevation between the pump and distribution pipe invert plus the friction loss and not less than 2.5 feet (762 mm) where using low pressure distribution in the delivery pipe and network pipe.

705.2 Pump and alarm controls. The control system for the pumping chamber shall consist of a control for operating the pump and an alarm system to detect a pump. Pump start and stop depth controls shall be adjustable. Pump and alarm controls shall be of an approved type. Switches shall be resistant to sewage corrosion.

705.3 Alarm system. Alarm systems shall consist of a bell or light, mounted in the structure, and shall be located to be easily seen or heard. The high-water sensing device shall be installed approximately 2 inches (51 mm) above the depth set for the "on" pump control but below the bottom of the inlet to the pumping chamber. Alarm systems shall be installed on a separate circuit from the electrical service.

705.4 Electrical connections. Electrical connections shall be located outside the pumping chamber.

SECTION 706
DOSING

706.1 General. The dosing frequency shall be not greater than four times daily. A volume per dose shall be established by dividing the daily wastewater flow by the dosing frequency. The dosing volume shall be not less than 10 times the capacity of the distribution pipe volume. Table 706.1 provides the estimated volume for various pipe diameters.

TABLE 706.1
ESTIMATED VOLUME FOR VARIOUS DIAMETER PIPES

DIAMETER (inches)	VOLUME (gallons per foot length)
1	0.041
$1^1/_4$	0.064
$1^1/_2$	0.092
2	0.164
3	0.368
4	0.655
5	1.47

For SI: 1 inch = 25.4 mm, 1 gallon per foot = 0.012 L/mm.

CHAPTER 8

TANKS

User note:

About this chapter: *Septic tanks and other treatment tanks are key components of private sewage disposal systems as they must be properly sized to achieve the desired reduction of sewage to it basic components of sludge and nonpotable water. Chapter 8 covers sizing, capacity and installation of these tanks and water holding tanks.*

SECTION 801 GENERAL

801.1 Scope. The provisions of this chapter shall govern the design, installation, repair and maintenance of septic tanks, treatment tanks and holding tanks.

SECTION 802 SEPTIC TANKS AND OTHER TREATMENT TANKS

802.1 General. Septic tanks shall be fabricated or constructed of welded steel, monolithic concrete, fiberglass or an approved material. Tanks shall be water tight and fabricated to constitute an individual structure, and shall be designed and constructed to withstand anticipated loads. The design of prefabricated septic tanks shall be approved. Plans for site-constructed concrete tanks shall be approved prior to construction.

802.2 Design of septic tanks. Septic tanks shall have not less than two compartments. The inlet compartment shall be not less than two-thirds of the total capacity of the tank, not less than a 500-gallon (1893 L) liquid capacity and not less than 3 feet (914 mm) wide and 5 feet (1524 mm) long. The secondary compartment of a septic tank shall have not less than a capacity of 250 gallons (946 L) and not more than one-third of the total capacity. The secondary compartment of septic tanks having a capacity more than 1,500 gallons (5678 L) shall be not less than 5 feet (1524 mm) long.

The liquid depth shall be not less than 30 inches (762 mm) and a maximum average of 6 feet (1829 mm). The total depth shall be not less than 8 inches (203 mm) greater than the liquid depth.

Rectangular tanks shall be constructed with the longest dimensions parallel to the direction of the flow.

Cylindrical tanks shall be not less than 48 inches (1219 mm) in diameter.

802.3 Inlets and outlets. The inlet and outlet on all tanks or tank compartments shall be provided with open-end coated sanitary tees or baffles made of approved materials constructed to distribute flow and retain scum in the tank or compartments. The inlet and outlet openings on all tanks shall contain a stop or other provision that will prevent the insertion of the sewer piping beyond the inside wall of the tank. The tees or baffles shall extend not less than 6 inches (152 mm) above the liquid level, not less than 9 inches (229 mm) below the liquid level, and not greater than one-third the liquid depth below the liquid level. Not less than 2 inches (51 mm) of clear space shall be provided above the top of the baffles or tees. The bottom of the outlet opening shall be not less than 2 inches (51 mm) below the bottom of the inlet.

802.4 Manholes. Each compartment of a tank shall be provided with not fewer than one manhole opening located over the inlet or outlet opening, and such opening shall be not less than 24 inches (610 mm) square or 24 inches (610 mm) in diameter. Where the inlet compartment of a septic tank exceeds 12 feet (3658 mm) in length, an additional manhole shall be provided over the baffle wall. Manholes shall terminate not greater than 6 inches (152 mm) below the ground surface. Manholes shall be of the same material as the tank. Steel tanks shall have not less than a 2-inch (51 mm) collar for the manhole extensions permanently welded to the tank. The manhole extension on fiberglass tanks shall be of the same material as the tank and an integral part of the tank. The collar shall be not less than 2 inches (51 mm) high.

802.5 Manhole covers. Manhole risers shall be provided with a fitted, water-tight cover of concrete, steel, cast iron or other approved material capable of withstanding all anticipated loads. Manhole covers terminating above grade shall have an approved locking device.

802.6 Inspection opening. An inspection opening shall be provided over either the inlet or outlet baffle of every treatment tank. The opening shall be not less than 4 inches (102 mm) in diameter with a tight-fitting cover. Inspection pipes terminating above ground shall be not less than 6 inches (152 mm) above finished grade. Inspection pipes approved for terminating below grade shall be not more than 2 inches (51 mm) below finished grade, and the location shall be permanently recorded.

802.7 Capacity and sizing. The capacity of a septic tank or other treatment tank shall be based on the number of persons using the building to be served or on the volume and type of waste, whichever is greater. The minimum liquid capacity

TABLE 802.7.1
SEPTIC TANK CAPACITY
FOR ONE- AND TWO-FAMILY DWELLINGS

NUMBER OF BEDROOMS	SEPTIC TANK (gallons)
1	750
2	750
3	1,000
4	1,200
5	1,425
6	1,650
7	1,875
8	2,100

For SI: 1 gallon = 3.785 L.

shall be 750 gallons (2839 L). Where the required capacity is to be provided by more than one tank, the minimum capacity of any tank shall be 750 gallons (2839 L). The installation of more than four tanks in series is prohibited.

802.7.1 Sizing of tank. The minimum liquid capacity for one- and two-family dwellings shall be in accordance with Table 802.7.1.

802.7.2 Other buildings. For buildings, the liquid capacity shall be increased above the 750-gallon (2839 L) minimum as established in Table 802.7.1. In buildings with kitchen or laundry waste, the tank capacity shall be increased to receive the anticipated volume for a 24-hour period from the kitchen or laundry or both. The liquid capacities established in Table 802.7.2 do not include employees.

Exception: One- or two-family dwellings.

802.8 Installation. Septic and other treatment tanks shall be located with a horizontal distance not less than specified in Table 802.8 between various elements. Tanks installed in ground water shall be securely anchored. A 3-inch-thick (76 mm) compacted bedding shall be provided for all septic and other treatment tank installations. The bedding material shall be sand, gravel, granite, limerock or other noncorrosive materials of such size that the material passes through a 0.5-inch (12.7 mm) screen.

TABLE 802.8
MINIMUM HORIZONTAL SEPARATION DISTANCES FOR TREATMENT TANKS

ELEMENT	DISTANCE (feet)
Building	5
Cistern	25
Foundation wall	5
Lake, high water mark	25
Lot line	2
Pond	25
Reservoir	25
Spring	50
Stream or watercourse	25
Swimming pool	15
Water service	5
Well	25

For SI: 1 foot = 304.8 mm.

802.9 Backfill. The backfill material for steel and fiberglass tanks shall be specified for bedding and shall be tamped into place without causing damage to the coating. The backfill for concrete tanks shall be soil material, which shall pass a 4-inch (102 mm) screen and be tamped into place.

802.10 Manhole riser joints. Joints on concrete risers and manhole covers shall be tongue-and-groove or shiplap type and sealed water tight using neat cement, mortar or bituminous compound. Joints on steel risers shall be welded or flanged and bolted and water tight. Steel manhole extensions shall be bituminous coated both inside and outside. Methods of attaching fiberglass risers shall be water tight and approved.

802.11 Dosing or pumping chambers. Dosing or pumping chambers shall be fabricated or constructed of welded steel, monolithic concrete, glass fiber-reinforced polyester or other approved materials. Manholes for dosing or pumping chambers shall terminate not less than 4 inches (102 mm) above the ground surface. Dosing or pumping chambers shall be water tight, and materials and construction specifications shall meet the same criteria specified for septic tanks in this chapter.

802.11.1 Capacity sizing. The working capacity of the dosing or pumping chamber shall be sized to permit automatic discharge of the total daily sewage flow with discharge occurring not more than four times per 24 hours. Minimum capacity of a dosing chamber shall be 500 gallons (1893 L) and a space shall be provided between the bottom of the pump and floor of the dosing or pumping chamber. A dosing chamber shall have a 1-day holding capacity located above the high-water alarm for one- and two-family dwellings based on 100 gallons (379 L) per day per bedroom, or in the case of other buildings, in accordance with Section 802.7. Minimum pump chamber sizes are indicated for one- and two-family dwellings in Table 802.11.1. Where the total developed length of distribution piping exceeds 1,000 feet (305 m), the dosing or pumping chamber shall have two siphons or pumps dosing alternately and serving one-half of the soil absorption system.

TABLE 802.11.1
PUMP CHAMBER SIZES

NUMBER OF BEDROOMS	MINIMUM PUMPING CHAMBER SIZE (gallons)
1	500
2	500
3	750
4	750
5	1,000

For SI: 1 gallon = 3.785 L.

802.12 Design of other treatment tanks. The design of other treatment tanks shall be approved on an individual basis. The capacity, sizing and installation of the tank shall be in accordance with this section except as otherwise approved. Where a treatment tank is preceded by a conventional septic tank, credit shall be given for the capacity of the septic tank.

SECTION 803
MAINTENANCE AND SLUDGE DISPOSAL

803.1 Maintenance. Septic tanks and other treatment tanks shall be cleaned whenever the sludge and scum occupy one-third of the tank's liquid capacity.

803.2 Septage. Septage shall be disposed of at an approved location.

TABLE 802.7.2
ADDITIONAL CAPACITY FOR OTHER BUILDINGS

BUILDING CLASSIFICATION	CAPACITY (gallons)
Apartment buildings (per bedroom—includes automatic clothes washer)	150
Assembly halls (per person—no kitchen)	2
Bars and cocktail lounges (per patron space)	9
Beauty salons (per station—includes customers)	140
Bowling centers (per lane)	125
Bowling centers with bar (per lane)	225
Camp, day use only—no meals served (per person)	15
Campgrounds and camping resorts (per camp space)	100
Campground sanitary dump stations (per camp space) (omit camp spaces with sewer connection)	5
Camps, day and night (per person)	40
Car washes (per car handwash)	50
Catch basins—such as for garages and motor-fuel-dispensing facilities (per basin)	100
Catch basins—truck washing (per truck)	100
Places of religious worship—no kitchen (per person)	3
Places of religious worship—with kitchen (per person)	7.5
Condominiums (per bedroom—includes automatic clothes washer)	150
Dance halls (per person)	3
Dining halls—kitchen and toilet waste—with dishwasher, food waste grinder or both (per meal served)	11
Dining halls—kitchen waste only (per meal served)	3
Drive-in restaurants—all paper service (per car space)	15
Drive-in restaurants—all paper service, inside seating (per seat)	15
Drive-in theaters (per car space)	5
Employees—in all buildings, per employee—total all shifts	20
Floor drains (per drain)	50
Hospitals (per bed space)	200
Hotels or motels and tourist rooming houses	100
Labor camps, central bathhouses (per employee)	30
Medical office buildings, clinics and dental offices Doctors, nurses, medical staff (per person) Office personnel (per person) Patients (per person)	 75 20 10
Mobile home parks, homes with bathroom groups (per site)	300
Motor-fuel-dispensing facilities	10
Nursing and rest homes—without laundry (per bed space)	100
Outdoor sports facilities (toilet waste only—per person)	5
Parks, toilet waste (per person—75 persons per acre)	5
Parks, with showers and toilet waste (per person—75 persons per acre)	10
Restaurants—dishwasher or food waste grinder or both (per seat)	3
Restaurants—kitchen and toilet wastes (per seating space)	30
Restaurants—kitchen waste only—without dishwasher and food waste grinder (per seat)	9
Restaurants—toilet waste only (per seat)	21
Restaurants (24-hour)—dishwasher or food waste grinder (per seat)	6

(continued)

**TABLE 802.7.2—continued
ADDITIONAL CAPACITY FOR OTHER BUILDINGS**

BUILDING CLASSIFICATION	CAPACITY (gallons)
Restaurants (24 hour)—kitchen and toilet waste (per seating space)	60
Retail stores—customers	1.5
Schools (per classroom—25 pupils per classroom)	450
Schools with meals served (per classroom—25 pupils per classroom)	600
Schools with meals served and showers provided (per classroom)	750
Self-service laundries (toilet waste only, per machine) Automatic clothes washers—such as for apartments and service buildings, (per machine)	50 300
Showers—public (per shower taken)	15
Swimming pool bathhouses (per person)	10

For SI: 1 gallon = 3.785 L.

SECTION 804 CHEMICAL RESTORATION

804.1 General. Products for chemical restoration or chemical restoration procedures for *private sewage disposal systems* shall not be used unless approved.

SECTION 805 HOLDING TANKS

805.1 Approval. The installation of a holding tank shall not be approved where the site can accommodate the installation of any other *private sewage disposal system* specified in this code. A pumping and maintenance schedule for each holding tank installation shall be submitted to the code official.

805.2 Sizing. The minimum liquid capacity of a holding tank for one- and two-family dwellings shall be in accordance with Table 805.2. Other buildings shall have a minimum 5-day holding capacity, but not less than 2,000 gallons (7570 L). Sizing shall be in accordance with Table 802.7.2. Not more than four holding tanks shall be installed in series.

**TABLE 805.2
MINIMUM LIQUID CAPACITY OF HOLDING TANKS**

NUMBER OF BEDROOMS	TANK CAPACITY (gallons)
1	2,000
2	2,000
3	2,000
4	2,500
5	3,000
6	3,500
7	4,000
8	4,500

For SI: 1 gallon = 3.785 L.

805.3 Construction. Holding tanks shall be constructed of welded steel, monolithic concrete, glass-fiber-reinforced polyester or other approved materials.

805.4 Installation. Tanks shall be located in accordance with Section 802.8, except the tanks shall be not less than 20 feet (6096 mm) from any part of a building. Holding tanks shall be located so the servicing manhole is located not less than 10 feet (3048 mm) from an all-weather access road or drive.

805.5 Warning device. A high-water warning device shall be installed to activate 1 foot (305 mm) below the inlet pipe. This device shall be either an audible or an approved illuminated alarm. The electrical junction box, including warning equipment junctions, shall be located outside the holding tank or housed in waterproof, explosionproof enclosures. Electrical relays or controls shall be located outside the holding tank.

805.6 Manholes. Each tank shall be provided with either a manhole not less than 24 inches (610 mm) square or with a manhole having a 24-inch (610 mm) inside diameter extending not less than 4 inches (102 mm) above ground. Finished grade shall be sloped away from the manhole to divert surface water from the manhole. Each manhole cover shall have an effective locking device. Service ports in manhole covers shall be not less than 8 inches (203 mm) in diameter and shall be 4 inches (102 mm) above finished grade level. The service port shall have an effective locking cover or a brass cleanout plug.

805.7 Septic tank. The outlet shall be sealed where an approved septic tank is installed to serve as a holding tank. Removal of the inlet and outlet baffle shall not be prohibited.

805.8 Vent. Each tank shall be provided with a vent not less than 2 inches (51 mm) in diameter and shall extend not less than 12 inches (305 mm) above finished grade, terminating with a return bend fitting or approved vent cap.

CHAPTER 9
MOUND SYSTEMS

User note:

About this chapter: *Mound systems are another method for disposal of the nonpotable water from a septic tank. The existing soil at the site may not have the capacity to absorb the water because of seasonal ground water conditions or presence of bedrock just below the soil surface. Chapter 9 provides extensive design information and installation methods to build mound systems for absorbing the nonpotable water.*

SECTION 901
GENERAL

901.1 Scope. The provisions of this chapter shall govern the design and installation of mound systems.

SECTION 902
SOIL AND SITE REQUIREMENTS

902.1 Soil borings. Not less than three *soil borings* per site shall be conducted in accordance with Chapter 4 to determine the depth to seasonal or permanent soil saturation or bedrock. Identification of a replacement system area is not required.

902.2 Prohibited locations. A mound system shall be prohibited on sites not having the minimum depths of soil specified in Table 902.2. The installation of a mound in a filled area shall be prohibited. A mound shall not be installed in a compacted area or over a failing conventional system.

TABLE 902.2
MINIMUM SOIL DEPTHS FOR MOUND SYSTEM INSTALLATION

RESTRICTING FACTOR	MINIMUM SOIL DEPTH TO RESTRICTION (inches)
High ground water	24
Impermeable rock strata	60
Pervious rock	24
Rock fragments (50-percent volume)	24

For SI: 1 inch = 25.4 mm.

902.3 Slowly permeable soils with or without high ground water. Percolation tests shall be conducted at a depth of 20 inches to 24 inches (508 mm to 610 mm) from existing grade. Where a more slowly permeable horizon exists at less than 20 inches to 24 inches (508 mm to 610 mm), percolation tests shall be conducted within that horizon. A mound system shall be suitable for such site condition where the percolation rate is greater than 60 minutes per inch and less than or equal to 120 minutes per inch (2.4 min/mm to 4.7 min/mm).

902.4 Shallow permeable soils over creviced bedrock. Percolation tests shall be conducted at a depth of 12 inches to 18 inches (305 mm to 457 mm) from existing grade. Where a more slowly permeable horizon exists within 12 inches to 18 inches (305 mm to 457 mm), percolation tests shall be conducted within that horizon. A mound system shall be suitable for such site condition where the percolation rate is between 3 minutes per inch and 60 minutes per inch (0.12 min/mm and 2.4 min/mm).

902.5 Permeable soils with high ground water. Percolation tests shall be conducted at a depth of 20 inches to 24 inches (508 mm to 610 mm) from existing grade. Where a more slowly permeable horizon exists at less than 20 inches to 24 inches (508 mm to 610 mm), percolation tests shall be conducted within that horizon. A mound system shall be suitable for such site condition where the percolation rate is between 0 minutes per inch and 60 minutes per inch (0 min/mm and 2.4 min/mm).

902.6 Depth to pervious rock. Not less than 24 inches (610 mm) of unsaturated natural soil shall be over creviced or porous bedrock.

902.7 Depth to high ground water. Not less than 24 inches (610 mm) of unsaturated natural soil shall be present over high ground water as indicated by soil mottling or direct observation of water in accordance with Chapter 4.

902.8 Slopes. A mound shall not be installed on a slope greater than 6 percent where the percolation rate is between 30 and 120 minutes per inch (1.2 and 4.7 min/mm). The maximum allowable slope shall be 12 percent where there is a complex slope (two directions).

902.9 Location of mound on sloping sites. The mound shall be located so the longest dimension of the mound and the distribution lines are perpendicular to the slope. The mound shall be placed upslope and not at the base of a slope. The mound shall be situated so the effluent is not concentrated in one direction where there is a complex slope (two directions). Surface water runoff shall be diverted around the mound.

902.10 Depth to rock strata or 50 percent by volume rock fragments. Not less than 60 inches (1524 mm) of soil shall be present over uncreviced, impermeable bedrock. Where the soil contains 50-percent coarse fragments by volume in the upper 24 inches (610 mm), a mound shall not be installed except where there is not less than 24 inches (610 mm) of permeable, unsaturated soil with less than 50-percent coarse fragments located beneath this layer.

SECTION 903
SYSTEM DESIGN

903.1 Mound dimensions and design. For one- and two-family dwellings and other buildings with estimated wastewater flows less than 600 gallons (2271 L) per day, the mound dimensions shall be determined in accordance with this section or Tables 903.1(1) through 903.1(12). Dimensions and corresponding letter designations listed in the tables and referenced in this section are shown in Appendix A, Figures A-6 through A-10. For buildings with estimated wastewater flows exceeding 600 gallons (2271 L) per day, the mound shall be designed in accordance with this section.

Daily wastewater flow shall be estimated as 150 gallons (568 L) per day per bedroom for one- and two-family dwellings. For other buildings the total daily wastewater flow shall be determined in accordance with Table 802.7.2.

903.1.1 Symbols. The following symbols and notations shall apply to the provisions of this section.

A = Bed or trench width, feet (mm).

A_A = Required absorption area, square feet (m^2).

B = Bed or trench length, feet (mm).

B_A = Basal area, square feet (m^2).

C = Trench spacing, feet (mm).

C_I = Infiltration capacity of natural soil, gallons per foot per day (L/mm/day).

D = Fill depth, feet (mm).

E = Downslope fill depth, feet (mm).

F = Bed or trench depth, feet (mm).

G = Minimum cap and topsoil depth, feet (mm).

H = Cap and topsoil depth at center of mound, foot (mm).

I = Downslope width, feet (mm).

J = Upslope width, feet (mm).

K = End slope length, feet (mm).

L = Total mound length, feet (mm).

N = Number of trenches.

P = Distribution pipe length, feet (mm).

R = Manifold length, feet (mm).

S = Distribution pipe spacing, feet (mm).

S_D = Downslope correction factor.

S_U = Upslope correction factor.

T_W = Total daily wastewater flow, gallons per day (L/day).

W = Total mound width, feet (mm).

X = Slope, percent.

903.2 Size of absorption area. The absorption area shall be sized based on the daily wastewater flow and the infiltrative capacity of the medium sand texture fill material, equaling 1.2 gallons per square foot (0.03 L/m^2) per day. The required absorption area shall be determined by the following equation:

$$A_A = \frac{T_W}{1.2 \text{ gal./ft}^2\text{/day}} \qquad \textbf{(Equation 9-1)}$$

For SI: 1 square foot = 0.0929 m^2, 1 gallon = 3.785 L.

903.3 Trenches. Effluent shall be distributed in the mound through a trench system for slowly permeable soils with or without high ground water. Trench length shall be selected by determining the longest dimension perpendicular to any slope on the site. Trench width and spacing is dependent on specific site conditions. Trenches shall be 2 feet to 4 feet (610 mm to 1219 mm) wide. Trench length (B) shall be not more than 100 feet (2540 mm). Trenches shall be of equal length where more than one trench is required. A mound shall not have more than three trenches. Trench spacing (C) shall be determined by the following equation:

$$C = \frac{T_W}{N \times 0.24 \text{ gal./ft}^2\text{/day} \times B} \qquad \textbf{(Equation 9-2)}$$

For SI: 1 gallon = 3.785 L, 1 square foot = 0.0929 m^2.

The calculated trench spacing (C) shall be measured from center to center of the trenches. Facilities with more than 1,500 gallons (56 775 L) per day shall be specifically engineered and approved for use with a trench system.

903.4 Beds. A long, narrow bed design shall be used for permeable soils with high water tables. The bed shall be square or rectangular for shallow permeable soils over bedrock. The bed length (B) shall be set after determining the longest dimension available and perpendicular to any slope on the site.

903.5 Mound dimensions. The mound height consists of the fill depth, bed or trench depth, the cap and topsoil depth.

903.5.1 Fill depth. The fill depth (D) shall be not less than 1 foot (305 mm) for slowly permeable soils and permeable soils with high water tables and not less than 2 feet (610 mm) of fill shall be required for shallow permeable soils over bedrock. Additional fill shall be placed at the downslope end of the bed or trench where the site is not level so the bottom of the bed or trench is level. The downslope fill depth for bed systems shall be determined by the following equation:

$$E = D + XA \qquad \textbf{(Equation 9-3)}$$

For SI: 1 foot = 304.8 mm.

The downslope fill depth for trench systems shall be determined by the following equation:

$$E = D + X(C + A) \qquad \textbf{(Equation 9-4)}$$

For SI: 1 foot = 304.8 mm.

903.5.2 Bed or trench depth. The bed or trench depth (F) shall be not less than 9 inches (229 mm) and not less than 6 inches (152 mm) of aggregate shall be placed under the distribution pipes and not less than 2 inches (51 mm) of aggregate shall be placed over the top of the distribution pipes.

903.5.3 Cap and topsoil depth. The cap and topsoil depth (H) at the center of the mound shall be not less than 18 inches (457 mm), which includes 1 foot (305 mm) of subsoil and 6 inches (152 mm) of topsoil. Outer edges of the mound, G (the minimum cap and topsoil depth), shall be not less than 1 foot (305 mm), which includes 6 inches (152 mm) of subsoil and 6 inches (152 mm) of topsoil. The soil used for the cap shall be topsoil or finer textured subsoil.

903.5.4 Mound lengths. The total mound length (L) shall be determined by the following equation:

$$L = B + 2K \qquad \textbf{(Equation 9-5)}$$

For SI: 1 foot = 304.8 mm.

where:

$$K = 3\left[\frac{(D + E)}{2} + F + H\right]$$

TABLE 903.1(1)
DESIGN CRITERIA FOR A MOUND FOR A ONE-BEDROOM HOME ON A 0- TO 6-PERCENT SLOPE WITH LOADING RATES OF 150 GALLONS PER DAY FOR SLOWLY PERMEABLE SOIL

DESIGN PARAMETER		SLOPE (percent)			
		0	2	4	6
A	Trench width, feet	3	3	3	3
B	Trench length, feet Number of trenches	42 1	42 1	42 1	42 1
D	Mound height, inches	12	12	12	12
F	Mound height, inches	9	9	9	9
G	Mound height, inches	12	12	12	12
H	Mound height, inches	18	18	18	18
I	Mound width, feet[a]	15	15	15	15
J	Mound width, feet[a]	11	8	8	8
K	Mound length, feet	10	10	10	10
L	Mound length, feet	62	62	62	62
P	Distribution pipe length, feet Distribution pipe diameter, inches Number of holes per distribution pipe[b] Hole spacing, inches[b] Hole diameter, inches[b]	20 1 9 30 0.25	20 1 9 30 0.25	20 1 9 30 0.25	20 1 9 30 0.25
W	Mound width, feet	25	26	26	26

For SI: 1 inch = 25.4 mm, 1 foot = 304.8 mm, 1 gallon = 3.785 L.
a. Additional width to obtain required basal area.
b. Last hole is located at the end of the distribution pipe, which is 15 inches from the other hole.

TABLE 903.1(2)
DESIGN CRITERIA FOR A TWO-BEDROOM HOME FOR A MOUND ON A 0- TO 6-PERCENT SLOPE WITH LOADING RATES OF 300 GALLONS PER DAY FOR SLOWLY PERMEABLE SOIL

DESIGN PARAMETER		SLOPE (percent)			
		0	2	4	6
A	Trench width, feet	3	3	3	3
B	Trench length, feet Number of trenches	42 2	42 2	42 2	42 2
C	Trench spacing, feet	15	15	15	15
D	Mound height, inches	12	12	12	12
E	Mound height, inches	12	17	25	25
F	Mound height, inches	9	9	9	9
G	Mound height, inches	12	12	12	12
H	Mound height, inches	18	18	18	18
I	Mound width, feet[a]	12	20	20	20
J	Mound width, feet	12	8	8	8
K	Mound length, feet	10	10	10	10
L	Mound length, feet	62	62	62	62
P	Distribution pipe length, feet Distribution pipe diameter, inches Number of holes per distribution pipe[b] Hole spacing, inches[b] Hole diameter, inches	20 1 9 30 0.25	20 1 9 30 0.25	20 1 9 30 0.25	20 1 9 30 0.25
R	Manifold length, feet Manifold diameter, inches[c]	15 2	15 2	15 2	15 2
W	Mound width, feet	42	46	46	46

For SI: 1 inch = 25.4 mm, 1 foot = 304.8 mm, 1 gallon = 3.785 L.
a. Additional width to obtain required basal area.
b. Last hole is located at the end of the distribution pipe, which is 15 inches from the other hole.
c. Diameter dependent on the size of pipe from pump and inlet position.

TABLE 903.1(3)
DESIGN CRITERIA FOR A THREE-BEDROOM HOME FOR A MOUND ON A 0- TO 6-PERCENT SLOPE WITH LOADING RATES OF 450 GALLONS PER DAY FOR SLOWLY PERMEABLE SOIL

DESIGN PARAMETER		SLOPE (percent)			
		0	2	4	6
A	Trench width, feet	3	3	3	3
B	Trench length, feet Number of trenches	63 2	63 2	63 2	63 2
C	Trench spacing, feet	15	15	15	15
D	Mound height, inches	12	12	12	12
E	Mound height, inches	12	17	20	25
F	Mound height, inches	9	9	9	9
G	Mound height, inches	12	12	12	12
H	Mound height, inches	18	18	18	18
I	Mound width, feet[a]	12	20	20	20
J	Mound width, feet[a]	12	8	8	8
K	Mound length, feet	10	10	10	10
L	Mound length, feet	62	62	62	62
P	Distribution pipe length, feet Distribution pipe diameter, inches Number of holes per distribution pipe[b] Hole spacing, inches[b] Hole diameter, inches	31 $1^1/_4$ 13 30 0.25	31 $1^1/_4$ 13 30 0.25	31 $1^1/_4$ 13 30 0.25	31 $1^1/_4$ 13 30 0.25
R	Manifold length, feet Manifold diameter, inches[c]	15 2	15 2	15 2	15 2
W	Mound width, feet	42	46	46	46

For SI: 1 inch = 25.4 mm, 1 foot = 304.8 mm, 1 gallon = 3.785 L.

a. Additional width to obtain required basal area.

b. First hole is located 12 inches from the manifold.

c. Diameter dependent on the size of pipe from pump and inlet position.

TABLE 903.1(4)
DESIGN CRITERIA FOR A FOUR-BEDROOM HOME FOR A MOUND ON A 0- TO 6-PERCENT SLOPE WITH LOADING RATES OF 600 GALLONS PER DAY FOR SLOWLY PERMEABLE SOIL

DESIGN PARAMETER		SLOPE (percent)			
		0	2	4	6
A	Trench width, feet	3	3	3	3
B	Trench length, feet Number of trenches	56 3	56 3	56 3	56 3
C	Trench spacing, feet	15	15	15	15
D	Mound height, inches	12	12	12	12
E	Mound height, inches	12	20	28	36
F	Mound height, inches	9	9	9	9
G	Mound height, inches	12	12	12	12
H	Mound height, inches	24	24	24	24
I	Mound width, feet[a]	12	20	20	20
J	Mound width, feet[a]	12	8	8	8
K	Mound length, feet	12	12	12	14
L	Mound length, feet	80	80	80	84
P	Distribution pipe length, feet Distribution pipe diameter, inches Number of holes per distribution pipe[b] Hole spacing, inches[b] Hole diameter, inches	27.5 $1^1/_4$ 12 30 0.25	27.5 $1^1/_4$ 12 30 0.25	27.5 $1^1/_4$ 12 30 0.25	27.5 $1^1/_4$ 12 30 0.25
R	Manifold length, feet Manifold diameter, inches[c]	30 2	30 2	30 2	30 2
W	Mound width, feet	57	61	61	61

For SI: 1 inch = 25.4 mm, 1 foot = 304.8 mm, 1 gallon = 3.785 L.
a. Additional width to obtain required basal area.
b. Last hole is located at the end of the distribution pipe, which is 15 inches from the previous hole.
c. Diameter dependent on the size of pipe from pump and inlet position.

TABLE 903.1(5)
DESIGN CRITERIA FOR A ONE-BEDROOM HOME FOR A MOUND ON A 0- TO 12-PERCENT SLOPE WITH LOADING RATES OF 150 GALLONS PER DAY FOR SHALLOW PERMEABLE SOIL OVER CREVICED BEDROCK

DESIGN PARAMETER		PERCOLATION RATE (minutes per inch) SLOPE (percent)						
		3 to 60				3 to less than 30		
		0	2	4	6	8	10[a]	12[a]
A	Bed width, feet[b]	10	10	10	10	10	10	10
B	Bed length, feet	13	13	13	13	13	13	13
D	Mound height, inches	24	24	24	24	24	24	24
E	Mound height, inches	24	26	29	31	34	36	38
F	Mound height, inches	9	9	9	9	9	9	9
G	Mound height, inches	12	12	12	12	12	12	12
H	Mound height, inches	18	18	18	18	18	18	18
I	Mound width, feet	12	13	14	17	18	21	26
J	Mound width, feet	12	11	10	10	9	9	9
K	Mound length, feet	12	12	12	13	13	13	15
L	Mound length, feet	37	37	37	39	39	39	43

(continued)

TABLE 903.1(5)—continued
DESIGN CRITERIA FOR A ONE-BEDROOM HOME FOR A MOUND ON A 0- TO 12-PERCENT SLOPE WITH LOADING RATES OF 150 GALLONS PER DAY FOR SHALLOW PERMEABLE SOIL OVER CREVICED BEDROCK

DESIGN PARAMETER		PERCOLATION RATE (minutes per inch) SLOPE (percent)						
		3 to 60				3 to less than 30		
		0	2	4	6	8	10[a]	12[a]
P	Distribution pipe length, feet[c] Distribution pipe diameter, inches Number of distribution pipes	12.5 1 6	12.5 1 6	12.5 1 6	12.5 1 6	12.5 1 6	12.5 1 6	12.5 1 6
R	Manifold length, feet Manifold diameter, inches[c]	6 2	6 2	6 2	6 2	6 2	6 2	6 2
S	Distribution pipe spacing, feet Number of holes per distribution pipe[d] Hole spacing, inches[d] Hole diameter, inches	3 6 30 0.25	3 6 30 0.25	3 6 30 0.25	3 6 30 0.25	3 6 30 0.25	3 6 30 0.25	3 6 30 0.25
W	Mound width, feet	34	34	34	37	37	41	45

For SI: 1 inch = 25.4 mm, 1 foot = 304.8 mm, 1 gallon = 3.785 L, 1 minute per inch = 2.4 s/mm.

a. On sites with a 10- to 12-percent slope, the fill depth (*D*) shall be reduced to not less than 1.5 feet or the bed width shall be reduced to decrease *E* (downslope fill depth, feet).

b. Bed widths shall not be limited.

c. Use a manifold with distribution pipes on only one side.

d. Last hole is located at the end of the distribution pipe, which is 15 inches from the previous hole.

TABLE 903.1(6)
DESIGN CRITERIA FOR A TWO-BEDROOM HOME FOR A MOUND ON A 0- TO 12-PERCENT SLOPE WITH LOADING RATES OF 300 GALLONS PER DAY FOR SHALLOW PERMEABLE SOIL OVER CREVICED BEDROCK

DESIGN PARAMETER		PERCOLATION RATE (minutes per inch) SLOPE (percent)						
		3 to 60			3 to less than 30			
		0	2	4	6	8	10[a]	12[a]
A	Bed width, feet[b]	10	10	10	10	10	10	10
B	Bed length, feet	25	25	25	25	25	25	25
D	Mound height, inches	24	24	24	24	24	24	24
E	Mound height, inches	24	26	29	31	34	36	38
F	Mound height, inches	9	9	9	9	9	9	9
G	Mound height, inches	12	12	12	12	12	12	12
H	Mound height, inches	18	18	18	18	18	18	18
I	Mound width, feet	12	13	14	17	18	21	26
J	Mound width, feet	12	11	10	10	9	9	9
K	Mound length, feet	12	12	12	13	13	13	15
L	Mound length, feet	49	49	49	51	51	51	55
P	Distribution pipe length, feet[c] Distribution pipe diameter, inches Number of distribution pipes	12 1 6	12 1 6	12 1 6	12 1 6	12 1 6	12 1 6	12 1 6
R	Manifold length, feet Manifold diameter, inches	6 2	6 2	6 2	6 2	6 2	6 2	6 2
S	Distribution pipe spacing, feet Number of holes per distribution pipe[d] Hole spacing, inches[d] Hole diameter, inches	3 5 30 0.25	3 5 30 0.25	3 5 30 0.25	3 5 30 0.25	3 5 30 0.25	3 5 30 0.25	3 5 30 0.25
W	Mound width, feet	34	34	34	37	37	41	45

For SI: 1 inch = 25.4 mm, 1 foot = 304.8 mm, 1 gallon = 3.785 L, 1 minute per inch = 2.4 s/mm.

(continued)

TABLE 903.1(6)—continued
DESIGN CRITERIA FOR A TWO-BEDROOM HOME FOR A MOUND ON A 0- TO 12-PERCENT SLOPE WITH LOADING RATES OF 300 GALLONS PER DAY FOR SHALLOW PERMEABLE SOIL OVER CREVICED BEDROCK

a. On sites with a 10- to 12-percent slope, the fill depth (*D*) shall be reduced to not less than 1.5 feet or the bed width shall be reduced to decrease *E* (downslope fill depth, feet).

b. Bed widths shall not be limited.

c. This design is based on a manifold with distribution pipes on both sides. An alternative design basis is 24-foot distribution pipes, with manifold at the end.

d. Last hole is located 9 inches from the end of the distribution pipe.

TABLE 903.1(7)
DESIGN CRITERIA FOR A THREE-BEDROOM HOME FOR A MOUND ON A 0- TO 12-PERCENT SLOPE WITH LOADING RATES OF 450 GALLONS PER DAY FOR SHALLOW PERMEABLE SOIL OVER CREVICED BEDROCK

DESIGN PARAMETER		PERCOLATION RATE (minutes per inch) SLOPE (percent)						
		3 to 60		3 to less than 30				
		0	2	4	6	8	10[a]	12[a]
A	Bed width, feet[b]	10	10	10	10	10	10	10
B	Bed length, feet	38	38	38	38	38	38	38
D	Mound height, inches	24	24	24	24	24	24	24
E	Mound height, inches	24	26	29	31	34	36	38
F	Mound height, inches	9	9	9	9	9	9	9
G	Mound height, inches	12	12	12	12	12	12	12
H	Mound height, inches	18	18	18	18	18	18	18
I	Mound width, feet	12	13	14	17	18	21	26
J	Mound width, feet	12	11	10	10	9	9	9
K	Mound length, feet	12	12	12	13	13	13	15
L	Mound length, feet	62	62	62	64	64	64	68
P	Distribution pipe length, feet[c] Distribution pipe diameter, inches Number of distribution pipes	18.5 1 6	18.5 1 6	18.5 1 6	18.5 1 6	18.5 1 6	18.5 1 6	18.5 1 6
R	Manifold length, feet Manifold diameter, inches	6 2	6 2	6 2	6 2	6 2	6 2	6 2
S	Distribution pipe spacing, feet Number of holes per distribution pipe[d] Hole spacing, inches[d] Hole diameter, inches	3 8 30 0.25	3 8 30 0.25	3 8 30 0.25	3 8 30 0.25	3 8 30 0.25	3 8 30 0.25	3 8 30 0.25
W	Mound width, feet	34	34	34	37	37	41	45

For SI: 1 inch = 25.4 mm, 1 foot = 304.8 mm, 1 gallon = 3.785 L, 1 minute per inch = 2.4 s/mm.

a. On sites with a 10- to 12-percent slope, the fill depth (*D*) shall be reduced to not less than 1.5 feet or the bed width shall be reduced to decrease *E* (downslope fill depth, feet).

b. Bed widths shall not be limited.

c. Use a manifold with distribution pipes on only one side.

d. Last hole is located at the end of the distribution pipe, which is 27 inches from the previous hole.

TABLE 903.1(8)
DESIGN CRITERIA FOR A FOUR-BEDROOM HOME FOR A MOUND ON A 0- TO 12-PERCENT SLOPE WITH LOADING RATES OF 600 GALLONS PER DAY FOR SHALLOW PERMEABLE SOIL OVER CREVICED BEDROCK

DESIGN PARAMETER		PERCOLATION RATE (minutes per inch) SLOPE (percent)						
		3 to 60				3 to less than 30		
		0	2	4	6	8	10[a]	12[a]
A	Bed width, feet[b]	10	10	10	10	10	10	10
B	Bed length, feet	50	50	50	50	50	50	50
D	Mound height, inches	24	24	24	24	24	24	24
E	Mound height, inches	24	26	29	31	34	36	38
F	Mound height, inches	9	9	9	9	9	9	9
G	Mound height, inches	12	12	12	12	12	12	12
H	Mound height, inches	18	18	18	18	18	18	18
I	Mound width, feet	12	13	14	17	18	21	26
J	Mound width, feet	12	11	10	10	9	9	9
K	Mound length, feet	12	12	12	13	13	13	15
L	Mound length, feet	74	74	74	76	76	76	78
P	Distribution pipe length, feet[c] Distribution pipe diameter, inches Number of distribution pipes	24.5 1 6	24.5 1 6	24.5 1 6	24.5 1 6	24.5 1 6	24.5 1 6	24.5 1 6
R	Manifold length, feet Manifold diameter, inches	6 2	6 2	6 2	6 2	6 2	6 2	6 2
S	Distribution pipe spacing, feet Number of holes per distribution pipe[d] Hole spacing, inches[d] Hole diameter, inches	3 10 30 0.25	3 10 30 0.25	3 10 30 0.25	3 10 30 0.25	3 10 30 0.25	3 10 30 0.25	3 10 30 0.25
W	Mound width, feet	34	34	34	37	37	41	45

For SI: 1 inch 25.4 mm, 1 foot = 304.8 mm, 1 gallon = 3.785 L, 1 minute per inch = 2.4 s/mm.

a. On sites with a 10- to 12-percent slope, the fill depth (*D*) shall be reduced to not less than 1.5 feet or the bed width shall be reduced to decrease *E* (downslope fill depth, feet).

b. Bed widths shall not be limited.

c. Use a manifold with distribution pipes on only one side.

d. Last hole is located 9 inches from the end of the distribution pipe.

TABLE 903.1(9)
DESIGN CRITERIA FOR A ONE-BEDROOM HOME FOR A MOUND ON A 0- TO 12-PERCENT SLOPE WITH LOADING RATES OF 150 GALLONS PER DAY FOR PERMEABLE SOIL WITH A HIGH WATER TABLE

DESIGN PARAMETER		PERCOLATION RATE (minutes per inch) SLOPE (percent)						
		0 to 60				0 to less than 30		
		0	2	4	6	8	10	12
A	Bed width, feet	4	4	4	4	4	4	4
B	Bed length, feet	32	32	32	32	32	32	32
D	Mound height, inches	12	12	12	12	12	12	12
E	Mound height, inches	12	13	14	14	16	17	18
F	Mound height, inches	9	9	9	9	9	9	9
G	Mound height, inches	12	12	12	12	12	12	12
H	Mound height, inches	18	18	18	18	18	18	18
I	Mound width, feet	9	10	11	12	13	14	15
J	Mound width, feet	9	9	8	8	7	7	6

(continued)

TABLE 903.1(9)—continued
DESIGN CRITERIA FOR A ONE-BEDROOM HOME FOR A MOUND ON A 0- TO 12-PERCENT SLOPE WITH LOADING RATES OF 150 GALLONS PER DAY FOR PERMEABLE SOIL WITH A HIGH WATER TABLE

DESIGN PARAMETER		PERCOLATION RATE (minutes per inch) SLOPE (percent)						
		0 to 60				0 to less than 30		
		0	2	4	6	8	10	12
K	Mound length, feet	10	10	10	10	10	11	11
L	Mound length, feet	52	52	52	52	52	53	53
P	Distribution pipe length	15.5	15.5	15.5	15.5	15.5	15.5	15.5
	Distribution pipe diameter, inches	1	1	1	1	1	1	1
	Number of distribution pipes	2	2	2	2	2	2	2
	Number of holes per distribution pipe[a]	7	7	7	7	7	7	7
	Hole spacing, inches[a]	30	30	30	30	30	30	30
	Hole diameter, inches	0.25	0.25	0.25	0.25	0.25	0.25	0.25
W	Mound width, feet	22	23	23	24	24	25	25

For SI: 1 inch = 25.4 mm, 1 foot = 304.8 mm, 1 gallon = 3.785 L, 1 minute per inch = 2.4 s/mm.

a. Last hole is located at the end of the distribution pipe, which is 21 inches from the previous hole.

TABLE 903.1(10)
DESIGN CRITERIA FOR A TWO-BEDROOM HOME FOR A MOUND ON A 0- TO 12-PERCENT SLOPE WITH LOADING RATES OF 300 GALLONS PER DAY FOR PERMEABLE SOIL WITH A HIGH WATER TABLE

DESIGN PARAMETER		PERCOLATION RATE (minutes per inch) SLOPE (percent)						
		0 to 60				0 to less than 30		
		0	2	4	6	8	10	12
A	Bed width, feet	6	6	6	6	6	6	6
B	Bed length, feet	42	42	42	42	42	42	42
D	Mound height, inches	12	12	12	12	12	12	12
E	Mound height, inches	12	13	14	17	18	19	22
F	Mound height, inches	9	9	9	9	9	9	9
G	Mound height, inches	12	12	12	12	12	12	12
H	Mound height, inches	18	18	18	18	18	18	18
I	Mound width, feet	9	10	11	12	13	15	16
J	Mound width, feet	9	9	8	8	7	7	6
K	Mound length, feet	10	10	10	10	10	11	11
L	Mound length, feet	62	62	62	62	62	64	64
P	Distribution pipe length, feet[a]	20	20	20	20	20	20	20
	Distribution pipe diameter, inches	1	1	1	1	1	1	1
	Number of distribution pipes	4	4	4	4	4	4	4
R	Manifold length, feet	3	3	3	3	3	3	3
	Manifold diameter, inches	2	2	2	2	2	2	2
S	Distribution pipe spacing, feet	3	3	3	3	3	3	3
	Number of holes per distribution pipe[b]	9	9	9	9	9	9	9
	Hole spacing, inches[b]	30	30	30	30	30	30	30
	Hole diameter, inches	0.25	0.25	0.25	0.25	0.25	0.25	0.25
W	Mound width, feet	24	25	25	26	26	28	29

For SI: 1 inch = 25.4 mm, 1 foot = 304.8 mm, 1 gallon = 3.785 L, 1 minute per inch = 2.4 s/mm.

a. Use a manifold with distribution pipes only on one side.

b. Last hole is located at the end of the distribution pipe, which is 15 inches from the previous hole.

TABLE 903.1(11)
DESIGN CRITERIA FOR A THREE-BEDROOM HOME FOR A MOUND ON A 0- TO 12-PERCENT SLOPE WITH LOADING RATES OF 450 GALLONS PER DAY FOR PERMEABLE SOIL WITH A HIGH WATER TABLE

DESIGN PARAMETER		PERCOLATION RATE (minutes per inch) SLOPE (percent)						
		0 to 60				0 to less than 30		
		0	2	4	6	8	10	12
A	Bed width, feet	8	8	8	8	8	8	8
B	Bed length, feet	47	47	47	47	47	47	47
D	Mound height, inches	12	12	12	12	12	12	12
E	Mound height, inches	12	12	16	18	19	22	24
F	Mound height, inches	9	9	9	9	9	9	9
G	Mound height, inches	12	12	12	12	12	12	12
H	Mound height, inches	18	18	18	18	18	18	18
I	Mound width, feet	9	11	12	13	15	17	18
J	Mound width, feet	9	9	8	8	7	7	6
K	Mound length, feet	10	10	10	10	10	11	12
L	Mound length, feet	67	67	67	67	69	69	71
P	Distribution pipe length, feet Distribution pipe diameter, inches Number of distribution pipes	23 1 6	23 1 6	23 1 6	23 1 6	23 1 6	23 1 6	23 1 6
R	Manifold length, feet Manifold diameter, inches	64 2	64 2	64 2	64 2	64 2	64 2	64 2
S	Distribution pipe spacing, feet Number of holes per distribution pipe[a] Hole spacing, inches[a] Hole diameter, inches	32 10 30 0.25	32 10 30 0.25	32 10 30 0.25	32 10 30 0.25	32 10 30 0.25	32 10 30 0.25	32 10 30 0.25
W	Mound width, feet	26	28	28	29	30	32	32

For SI: 1 inch = 25.4 mm, 1 foot = 304.8 mm, 1 gallon = 3.785 L, 1 minute per inch = 2.4 s/mm.

a. Last hole is located at the end of the distribution pipe, which is 21 inches from the previous hole.

TABLE 903.1(12)
DESIGN CRITERIA FOR A FOUR-BEDROOM HOME FOR A MOUND ON A 0- TO 12-PERCENT SLOPE WITH LOADING RATES OF 600 GALLONS PER DAY FOR PERMEABLE SOIL WITH A HIGH WATER TABLE

DESIGN PARAMETER		PERCOLATION RATE (minutes per inch) SLOPE (percent)						
		0 to 60				0 to less than 30		
		0	2	4	6	8	10	12
A	Bed width, feet	10	10	10	10	10	10	10
B	Bed length, feet	50	50	50	50	50	50	50
D	Mound height, inches	12	12	12	12	12	12	12
E	Mound height, inches	12	14	17	19	22	24	26
F	Mound height, inches	9	9	9	9	9	9	9
G	Mound height, inches	12	12	12	12	12	12	12
H	Mound height, inches	18	18	18	18	18	18	18
I	Mound width, feet	9	11	13	14	17	18	19
J	Mound width, feet	9	9	8	8	7	7	6
K	Mound length, feet	10	10	10	10	11	11	12
L	Mound length, feet	70	70	70	70	72	72	74

(continued)

TABLE 903.1(12)—continued
DESIGN CRITERIA FOR A FOUR-BEDROOM HOME FOR A MOUND ON A 0- TO 12-PERCENT SLOPE WITH LOADING RATES OF 600 GALLONS PER DAY FOR PERMEABLE SOIL WITH A HIGH WATER TABLE

DESIGN PARAMETER		PERCOLATION RATE (minutes per inch) SLOPE (percent)						
		0 to 60				0 to less than 30		
		0	2	4	6	8	10	12
P	Distribution pipe length, feet	24.5	24.5	24.5	24.5	24.5	24.5	24.5
	Distribution pipe diameter, inches	1	1	1	1	1	1	1
	Number of distribution pipes	6	6	6	6	6	6	6
R	Manifold length, feet	6	6	6	6	6	6	6
	Manifold diameter, inches	2	2	2	2	2	2	2
S	Distribution pipe spacing, feet	3	3	3	3	3	3	3
	Number of holes per distribution pipe[a]	10	10	10	10	10	10	10
	Hole spacing, inches[a]	30	30	30	30	30	30	30
	Hole diameter, inches	0.25	0.25	0.25	0.25	0.25	0.25	0.25
W	Mound width, feet	28	29	31	32	34	35	36

For SI: 1 inch = 25.4 mm, 1 foot = 304.8 mm, 1 gallon = 3.785 L, 1 minute per inch = 2.4 s/mm.
a. Last hole is 9 inches from the end of the distribution pipe.

903.5.5 Mound widths. The mound width for a bed system shall be determined by the following equation:

$$W = J + A + I \qquad \textbf{(Equation 9-6)}$$

For SI: 1 foot = 304.8 mm.

The mound width for a trench system shall be determined by the following equation:

$$W = J + \frac{A}{2} + C(N-1) - \frac{A}{2} + I \qquad \textbf{(Equation 9-7)}$$

For SI: 1 foot = 304.8 mm.

where:

$J = 3(D + F + G)S_U$

$I = 3(E + F + G)S_D$

The upslope correction factor (S_U) and the downslope correction factor (S_D) shall be determined based on the slope in accordance with Table 903.5.5.

TABLE 903.5.5
DOWNSLOPE AND UPSLOPE WIDTH CORRECTIONS FOR MOUNDS ON SLOPING SITES

SLOPE (percent)	DOWNSLOPE CORRECTION FACTOR (S_D)	UPSLOPE CORRECTION FACTOR (S_U)
0	1	1
1	1.03	0.97
2	1.06	0.94
3	1.10	0.915
4	1.14	0.89
5	1.18	0.875
6	1.22	0.86
7	1.27	0.83
8	1.32	0.80
9	1.38	0.785
10	1.44	0.77
11	1.51	0.75
12	1.57	0.73

903.6 Basal area. The minimum basal area required shall be determined by the following equation:

$$B_A = \frac{T_W}{C_I} \qquad \textbf{(Equation 9-8)}$$

For SI: 1 square foot = 0.0929 m^2.

The infiltrative capacity of natural soil shall be determined on the percolation rate in accordance with Table 903.6.

TABLE 903.6
INFILTRATIVE CAPACITY OF NATURAL SOIL

PERCOLATION RATE (minutes per inch)	INFILTRATIVE CAPACITY (gallons per foot per day)
Less than 30	1.2
30 to 60	0.74
More than 60 to 120	0.24

For SI: 1 gallon per foot per day = 0.012 L/mm/day, 1 minute per inch = 2.4 s/mm.

903.6.1 Basal area available in bed system. The available basal area for a bed system shall be determined by one of the following equations:

$B_A = B(A + I)$ for sloping sites **(Equation 9-9)**

$B_A = BW$ for level sites **(Equation 9-10)**

For SI: 1 square foot = 0.0929 m^2.

903.6.2 Basal area available in trench system. The available basal area for a trench system shall be determined by one of the following equations:

$B_A = B\left(W + J + \frac{A}{2}\right)$ for sloping sites **(Equation 9-11)**

$B_A = BW$ for level sites **(Equation 9-12)**

For SI: 1 square foot = 0.0929 m^2.

903.6.3 Adequacy of basal area. The downslope width (I) on a sloping site shall be increased or the upslope width (J) and downslope (I) widths on a level site shall be increased until sufficient area is available if the basal area available is not equal to or greater than the basal area required.

903.7 Dose volume and pump. The dose volume and pump shall conform to the requirements of Chapters 7 and 8.

SECTION 904 CONSTRUCTION TECHNIQUES

904.1 General. Construction shall not commence where the soil is so wet a soil wire forms when the soil is rolled between the hands. Installation of mound systems where the soil on the site is frozen shall be prohibited for new construction.

904.2 Site preparation. Excess vegetation shall be cut and removed from the mound area. Small trees shall be cut to grade surface, leaving the stumps in place.

904.3 Force main. The force main from the pumping chamber shall be installed before the mound site is plowed. The force main shall be sloped uniformly toward the pumping chamber so the force main drains after each dose.

904.4 Plowing. The site shall be plowed with a moldboard plow or chisel plow. The site shall be plowed to a depth of 7 inches to 8 inches (178 mm to 203 mm) with the plowing perpendicular to the slope. Rototillers shall not be used. The sand fill shall be placed immediately after plowing. Foot and vehicular traffic shall be kept off the plowed area.

904.5 Sand fill material. The fill material shall be medium sand texture defined as 25 percent or more very coarse, coarse and medium sand and not more than 50 percent fine sand, very fine sand, silt and clay. The percentage of silt plus one and one-half times the percentage of clay shall not exceed 15 percent. Fill materials with higher content of silt and clay shall not be used.

904.5.1 Placement of sand fill. The medium sand fill shall be moved into place from the upslope and side edges of the plowed area. Vehicular traffic shall be prohibited in the area extending to 25 feet (7620 mm) beyond the downslope edge of the mound. The sand fill shall be moved into place with a track-type tractor and not less than 6 inches (152 mm) of sand shall be kept beneath the tracks at all times.

904.6 Installation of the absorption area. The bed or trenches shall be formed within the sand fill. The bottom of the trenches or bed shall be level. The elevation of the bottom of the trenches or bed shall be checked at the upslope and downslope edges to ensure that the fill has been placed to the proper depth.

904.7 Placement of the aggregate. Not less than 6 inches (152 mm) of coarse aggregate ranging in size from $^{1}/_{2}$ inch to $2^{1}/_{2}$ inches (12.7 mm to 64 mm) shall be placed in the bed or trench excavation. The top of the aggregate shall be level.

904.8 Distribution system. Distribution systems shall be placed on the aggregate, with the holes located on the bottom of the distribution pipe. The ends of all distribution pipes shall be marked at the surface, and an observation pipe shall be placed to the bottom of the bed or each trench.

904.9 Cover. The top of the bed or trenches shall be covered with not less than 2 inches (51 mm) of aggregate ranging in size from $^{1}/_{2}$ inch to $2^{1}/_{2}$ inches (12.7 mm to 64 mm) and not less than 4 inches to 5 inches (102 mm to 127 mm) of uncompacted straw or marsh hay or approved synthetic fabric shall be placed over the aggregate. Cap and topsoil covers shall be in place and the mound shall be seeded immediately and protected from erosion.

904.10 Maintenance. When the septic tank is pumped, the pump chamber shall be inspected and pumped to remove any solids present. Excess traffic in the mound area shall be avoided.

CHAPTER 10

CESSPOOLS

User note:

About this chapter: *Chapter 10 addresses the construction of cesspools. Although cesspools were a common method for disposing of sewage for new buildings in some areas of the United States, the code only allows the construction of a cesspool as a temporary measure until a proper sewage disposal system can be installed.*

SECTION 1001 GENERAL

1001.1 Scope. The provisions of this chapter shall govern the design and installation of *cesspools*.

1001.2 Application. *Cesspools* shall not be installed, except where approved by the code official. A *cesspool* shall be considered as only a temporary expedient pending the construction of a public sewer; as an overflow facility where installed in conjunction with an existing *cesspool*; or as a means of sewage disposal for limited, minor or temporary applications.

1001.3 Construction. *Cesspools* shall conform to the construction requirements of Section 605.3 for *seepage pits*. The *seepage pit* shall have a minimum sidewall of 20 feet (6096 mm) below the inlet opening. Where a stratum of gravel or equally pervious material of 4 feet (1219 mm) or more in thickness is found, the sidewall need not be more than 10 feet (3048 mm) below the inlet.

CHAPTER 11

RESIDENTIAL WASTEWATER SYSTEMS

User note:

About this chapter: *Septic tanks are not the only method for treatment of sewage from a residence. Chapter 11 indicates the standard with which factory-built wastewater treatment plants must comply.*

SECTION 1101
GENERAL

1101.1 Scope. The provisions of this chapter shall govern residential wastewater systems.

1101.2 Residential wastewater treatment systems. The regulations for materials, design, construction and performance shall comply with NSF 40.

CHAPTER 12

INSPECTIONS

User note:

About this chapter: *Chapter 12 covers the inspection requirements for all types of private sewage disposal systems.*

SECTION 1201 GENERAL

1201.1 Scope. The provisions of this chapter shall govern the inspection of *private sewage disposal systems*.

SECTION 1202 INSPECTIONS

1202.1 Initial inspection procedures. *Private sewage disposal systems* shall be inspected after construction, but before backfilling. The code official shall be notified when the *private sewage disposal system* is ready for inspection.

1202.2 Preparation for inspection. The installer shall make such arrangements as will enable the code official to inspect all parts of the system when a *private sewage disposal system* is ready. The installer shall provide the proper apparatus and equipment for conducting the inspection and furnish such assistance as is necessary to conduct the inspection.

1202.3 Covering of work. A *private sewage disposal system* or part thereof shall not be backfilled until such system has been inspected and approved. Any system that has been covered before being inspected and approved shall be uncovered as required by the code official.

1202.4 Other inspections. In addition to the required inspection prior to backfilling, the code official shall conduct any other inspections deemed necessary to determine compliance with this code.

1202.5 Inspections for additions, alterations or modifications. Additions, alterations or modifications to *private sewage disposal systems* shall be inspected.

1202.6 Defects in materials and workmanship. Where inspection discloses defective material, design or siting or unworkmanlike construction not conforming to the requirements of this code, the nonconforming parts shall be removed, replaced and reinspected.

CHAPTER 13
NONLIQUID SATURATED TREATMENT SYSTEMS

User note:

About this chapter: *Composting is another method for disposing of human waste. Chapter 13 references the standard that covers these nonliquid saturated treatment systems.*

SECTION 1301
GENERAL

1301.1 Scope. The provisions of this chapter shall govern nonliquid saturated treatment systems.

1301.2 Nonliquid saturated treatment systems. The regulations for materials, design, construction and performance shall comply with NSF 41.

CHAPTER 14
REFERENCED STANDARDS

User note:

About this chapter: *This code contains numerous references to standards promulgated by other organizations that are used to provide requirements for materials and methods of construction. Chapter 14 contains a comprehensive list of all standards that are referenced in this code. These standards, in essence, are part of this code to the extent of the reference to the standard.*

ASTM

ASTM International
100 Barr Harbor Drive, P.O. Box C700
West Conshohocken, PA 19428-2959

A74—15: Specification for Cast Iron Soil Pipe and Fittings
Table 505.1

A888—15: Specification for Hubless Cast Iron Soil Pipe and Fittings for Sanitary and Storm Drain, Waste, and Vent Piping Application
Table 505.1

B32—08(2014): Specification for Solder Metal
505.8.2

B75/B75M—11: Specification for Seamless Copper Tube
Table 505.1

B88—14: Specification for Seamless Copper Water Tube
Table 505.1

B251—10: Specification for General Requirements for Wrought Seamless Copper and Copper-alloy Tube
Table 505.1

B813—10: Specification for Liquid and Paste Fluxes for Soldering of Copper and Copper-alloy Tube
505.8.2

B828—02(2010): Practice for Making Capillary Joints by Soldering of Copper and Copper-alloy Tube and Fittings
505.8.2

C4—04(2014): Specification for Clay Drain Tile and Perforated Clay Drain Tile
Table 505.1

C14—15a: Specification for Nonreinforced Concrete Sewer, Storm Drain and Culvert Pipe
Table 505.1

C76—15A: Specification for Reinforced Concrete Culvert, Storm Drain and Sewer Pipe
Table 505.1

C425—04(2013): Specification for Compression Joints for Vitrified Clay Pipe and Fittings
505.12, 505.13

C428/C428M—05(2011)el: Specification for Asbestos-cement Nonpressure Sewer Pipe
Table 505.1

C443—12: Specification for Joints for Concrete Pipe and Manholes, Using Rubber Gaskets
505.7, 505.13

C564—14: Specification for Rubber Gaskets for Cast Iron Soil Pipe and Fittings
505.6.2, 505.6.3, 505.13

C700—13: Specification for Vitrified Clay Pipe, Extra Strength, Standard Strength and Perforated
Table 505.1

C913—08: Specification for Precast Concrete Water and Wastewater Structures
504.2

C1173—10(2014): Specification for Flexible Transition Couplings for Underground Piping Systems
505.3.1, 505.5.1, 505.7, 505.10.1, 505.12, 505.13

C1277—15: Specification for Shielding Coupling Joining Hubless Cast-iron Pipe and Fittings
505.6.3

ASTM—continued

C1440—08(2013): Specification for Thermoplastic Elastomeric (TPE) Gasket Materials for Drain, Waste and Vent (DWV), Sewer, Sanitary and Storm Plumbing Systems
505.13

C1460—2012: Specification for Shielded Transition Couplings for Use with Dissimilar DWV Pipe and Fittings Above Ground
505.13

C1461—08(2013): Specification for Mechanical Couplings Using Thermoplastic Elastomeric (TPE) Gaskets for Joining Drain, Waste and Vent (DWV) Sewer, Sanitary and Storm Plumbing Systems for Above and Below Ground Use
505.13

D1869—15: Specification for Rubber Rings for Fiber-Reinforced Cement Pipe
505.4, 505.13

D2235—04(2011): Specification for Solvent Cement for Acrylonitrile-butadiene-styrene (ABS) Plastic Pipe and Fittings
505.3.2, 505.5.2

D2564—12: Specification for Solvent Cements for Poly Vinyl Chloride (PVC) Plastic Piping Systems
505.10.2, 505.11.2

D2657—07: Standard Practice for Heat-fusion Joining of Polyolefin Pipe and Fittings
505.9.1

D2661—14: Specification for Acrylonitrile-butadiene-styrene (ABS) Schedule 40 Plastic Drain, Waste, and Vent Pipe and Fittings
Table 505.1, 505.3.2, 505.5.2

D2665—14: Specification for Poly Vinyl Chloride (PVC) Plastic Drain, Waste, and Vent Pipe and Fittings
Table 505.1

D2729—11: Specification for Poly Vinyl Chloride (PVC) Sewer Pipe and Fittings
Table 505.1.1

D2751—05: Specification for Acrylonitrile-butadiene-styrene (ABS) Sewer Pipe and Fittings
Table 505.1

D2855—96(2010): Standard Practice for Making Solvent-cemented Joints with Poly Vinyl Chloride (PVC) Pipe and Fittings
505.10.2, 505.11.2

D2949—10: Specification for 3.25-inch Outside Diameter Poly Vinyl Chloride (PVC) Plastic Drain, Waste, and Vent Pipe and Fittings
Table 505.1

D3034—14a: Specification for Type PSM Poly Vinyl Chloride (PVC) Sewer Pipe and Fittings
Table 505.1

D3212—07(2013): Specification for Joints for Drain and Sewer Plastic Pipes Using Flexible Elastomeric Seals
505.3.1, 505.5.1, 505.10.1

D4021—92: Specification for Glass-fiber Reinforced Polyester Underground Petroleum Storage Tanks
504.4

F405—05: Specification for Corrugated Polyethylene (PE) Pipe and Fittings
Table 505.1.1

F477—10: Specification for Elastomeric Seals (Gaskets) for Joining Plastic Pipe
505.13

F628—12E1: Specification for Acrylonitrile-butadiene-styrene (ABS) Schedule 40 Plastic Drain, Waste, and Vent Pipe with a Cellular Core
Table 505.1, 505.3.2, 505.5.2

F656—15: Specification for Primers for Use in Solvent Cement Joints of Poly Vinyl Chloride (PVC) Plastic Pipe and Fittings
505.10.2, 505.11.2

F891—10: Specification for Coextruded Poly Vinyl Chloride (PVC) Plastic Pipe with a Cellular Core
Table 505.1

F1488—14: Specification for Coextruded Composite Pipe
Table 505.1, Table 505.1.1

F1499—12: Specification for Coextruded Composite Drain Waste and Vent Pipe (DWV)
Table 505.1

CISPI

Cast Iron Soil Pipe Institute
2401 Fieldcrest Drive
Mundelein, IL 60060

301—12: Specification for Hubless Cast Iron Soil Pipe and Fittings for Sanitary and Storm Drain, Waste and Vent Piping Applications
Table 505.1

310—12: Specification for Coupling for Use in Connection with Hubless Cast Iron Soil Pipe and Fittings for Sanitary and Storm Drain, Waste and Vent Piping Applications
505.6.3

CSA

CSA Group
8501 East Pleasant Valley Road
Cleveland, OH 44131-5516

A257.2—14: Reinforced Circular Concrete Culvert, Storm Drain, Sewer Pipe and Fittings
Table 505.1

A257.3—14: Joints for Circular Concrete Sewer and Culvert Pipe, Manhole Sections and Fittings Using Rubber Gaskets
505.7, 505.13

B137.3—16: Rigid Poly Vinyl Chloride (PVC) Pipe for Pressure Applications
505.10.2, 505.11.2

B181.1—15: Acrylonitrile-butadiene-styrene (ABS) Drain, Waste, and Vent Pipe and Pipe Fittings
505.3.2, 505.5.2

B181.2—15: (PVC) Polyvinylchloride and Chlorinated Polyvinylchloride (CPVC) Drain, Waste, and Vent Pipe and Pipe Fittings
505.10.2, 505.11.2

B182.1—11: Plastic Drain and Sewer Pipe and Pipe Fittings
505.10.2, 505.11.2

B182.2—11: (PVC) Polyvinylchloride Sewer Pipe and Fittings PSM Type
Table 505.1

B182.4—15: Profile PVC Sewer Pipe and Fittings
Table 505.1

B602—15: Mechanical Couplings for Drain, Waste, and Vent Pipe and Sewer Pipe
505.3.1, 505.5.1, 505.6.3, 505 7, 505.10.1, 505.12, 505.13

CSA A257.1—14: Circular Concrete Culvert, Storm Drain, Sewer Pipe and Fittings
Table 505.1

ICC

International Code Council
500 New Jersey Avenue, NW
6th Floor
Washington, DC 20001

IBC—18: International Building Code®
201.3

IPC—18: International Plumbing Code®
201.3, 505.14

NSF

NSF International
789 N. Dixboro Road
P. O. Box 130140
Ann Arbor, MI 48105

40—2013: Residential Wastewater Treatment Systems
1102.1

41—2011: Nonliquid Saturated Treatment Systems (Composing Toilets)
1301.2

UL

UL LLC
333 Pfingsten Road
Northbrook, IL 60062-2096

70—2001: Septic Tanks, Bituminous-coated Metal
504.3

APPENDIX A

SYSTEM LAYOUT ILLUSTRATIONS

The provisions contained in this appendix are not mandatory unless specifically referenced in the adopting ordinance.

User note:

About this appendix: *Appendix A provides illustrations for many system layouts covered in some of the previous chapters.*

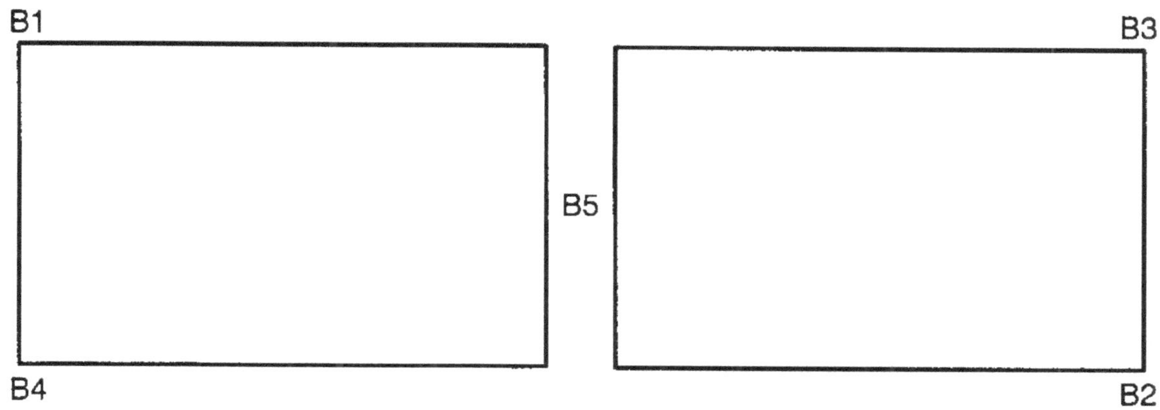

FIGURE A-1 (SECTION 403.1.1)
EXAMPLE OF SOIL-BORING LOCATIONS FOR TWO CONTIGUOUS ABSORPTION AREAS

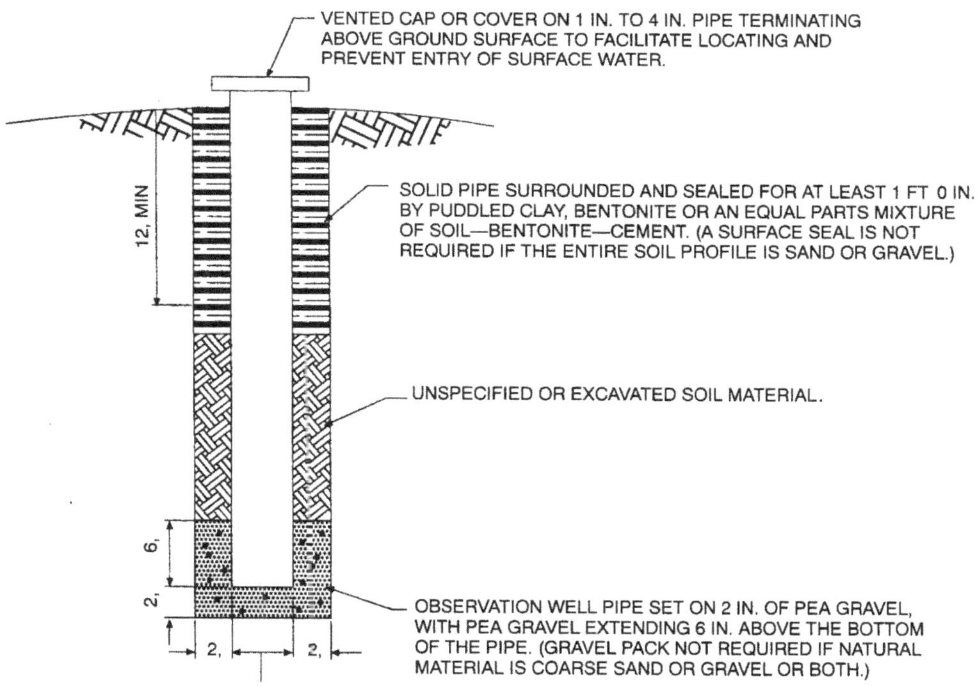

Note: Bore hole shall be 4 inches to 8 inches larger than the outside diameter of observation well pipe size.

For SI: 1 inch = 25.4 mm, 1 foot = 304.8 mm.

FIGURE A-2 (SECTION 405.2.4)
MONITORING WELL DESIGN

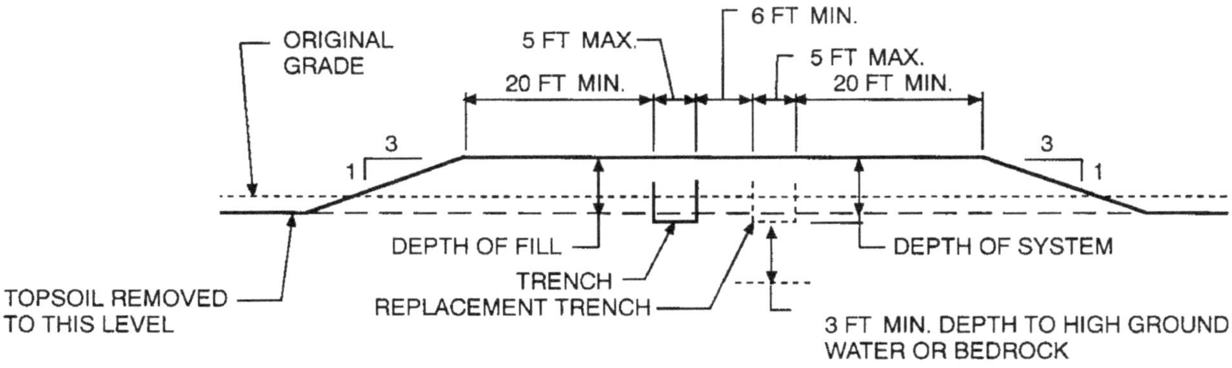

For SI: 1 foot = 304.8 mm.

FIGURE A-3 (SECTION 406.6.7)
DESIGN OF FILLED AREA SYSTEM

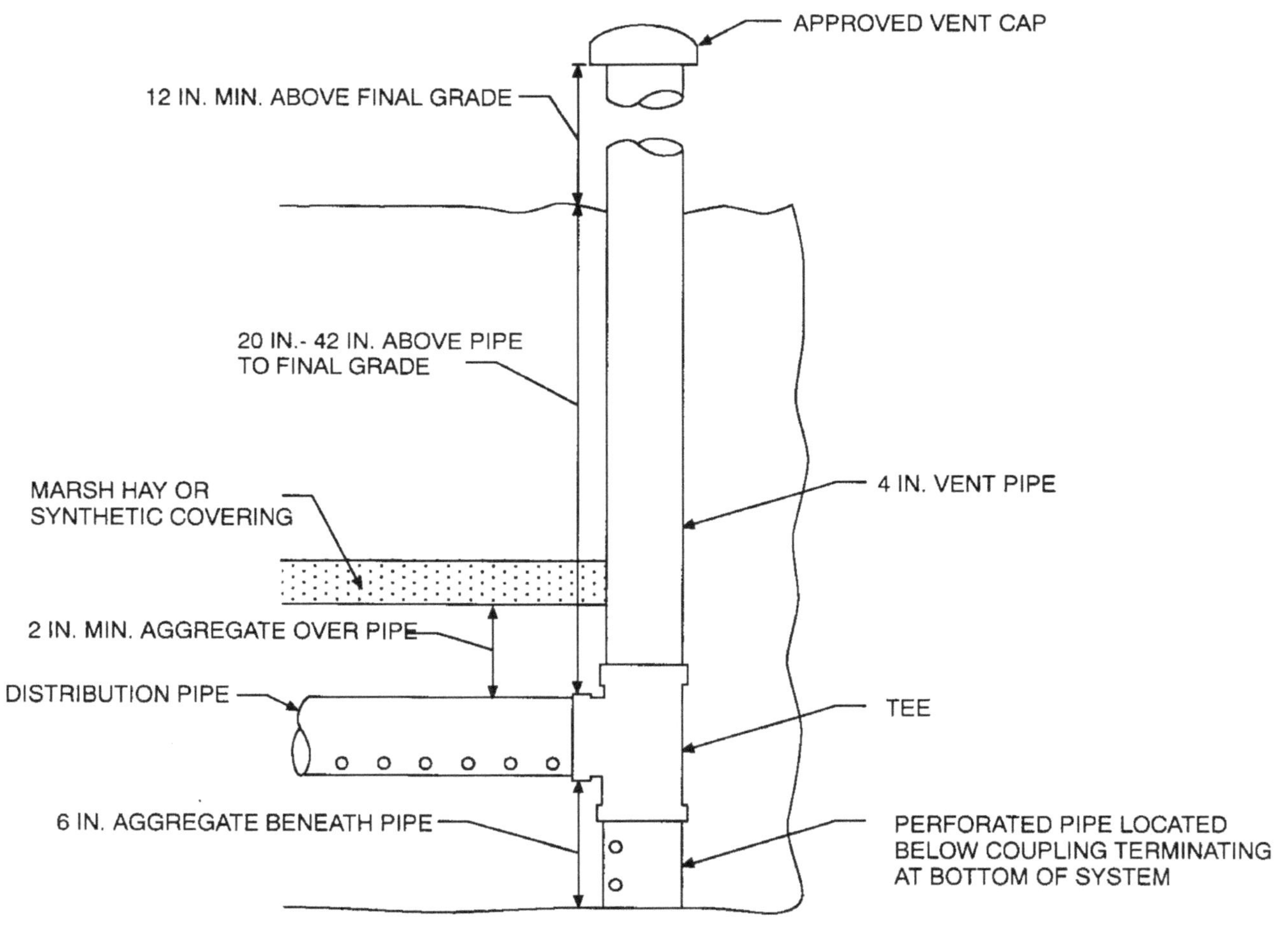

For SI: 1 inch = 25.4 mm.

FIGURE A-4 (SECTION 605.7)
OBSERVATION PIPE

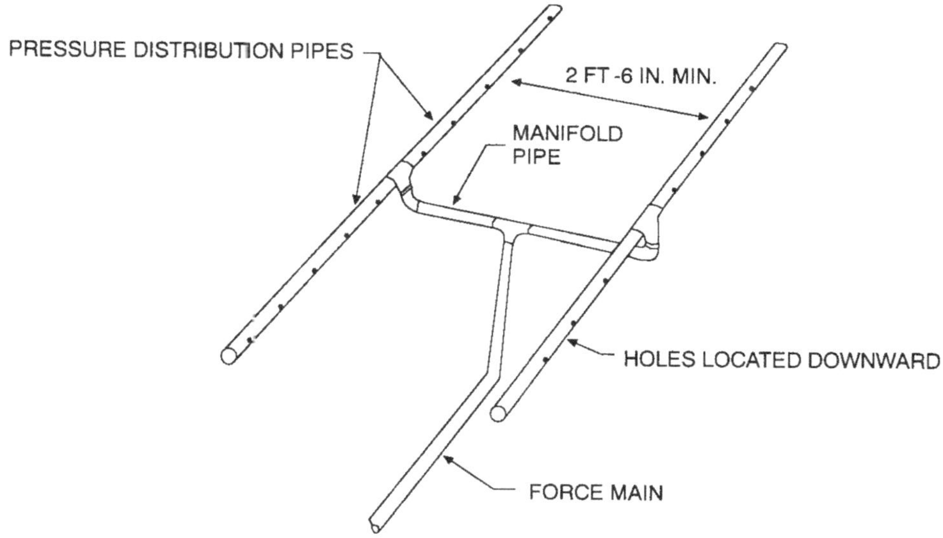

For SI: 1 inch = 25.4 mm, 1 foot = 304.8 mm.

FIGURE A-5 (SECTION 703.1)
PRESSURE DISTRIBUTION SYSTEM DESIGN

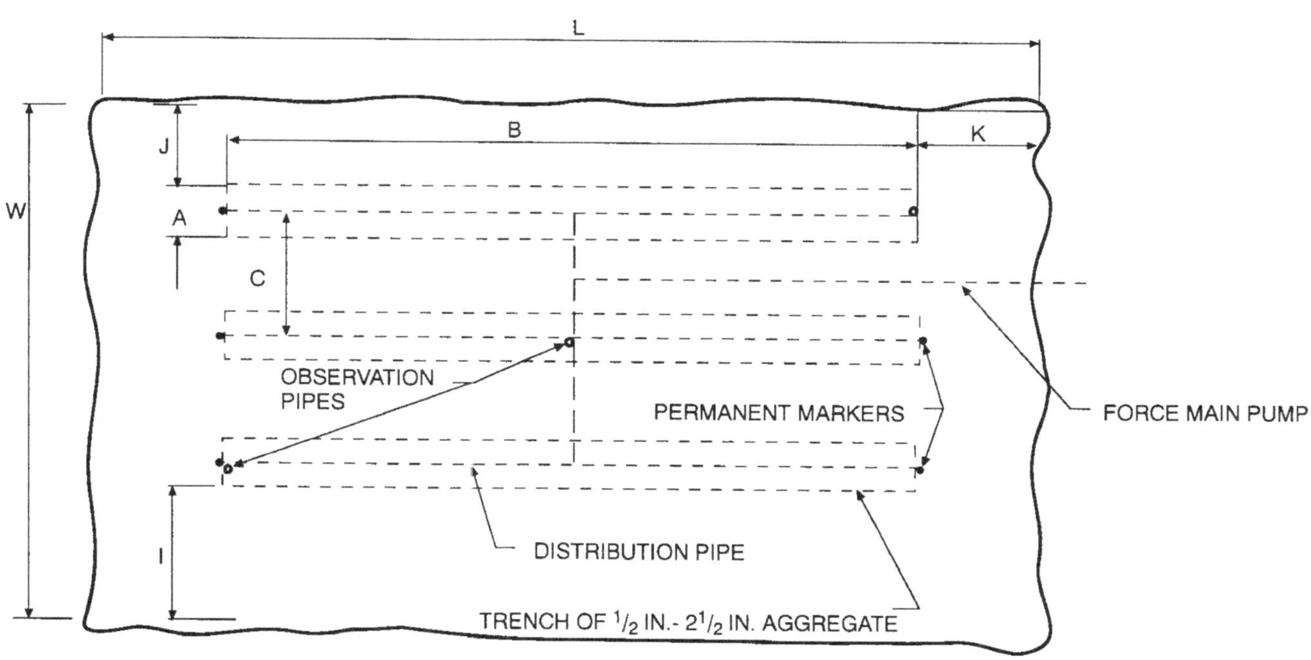

For SI: 1 inch = 25.4 mm.

FIGURE A-6 (SECTION 903.1)
MOUND USING THREE TRENCHES FOR ABSORPTION AREA

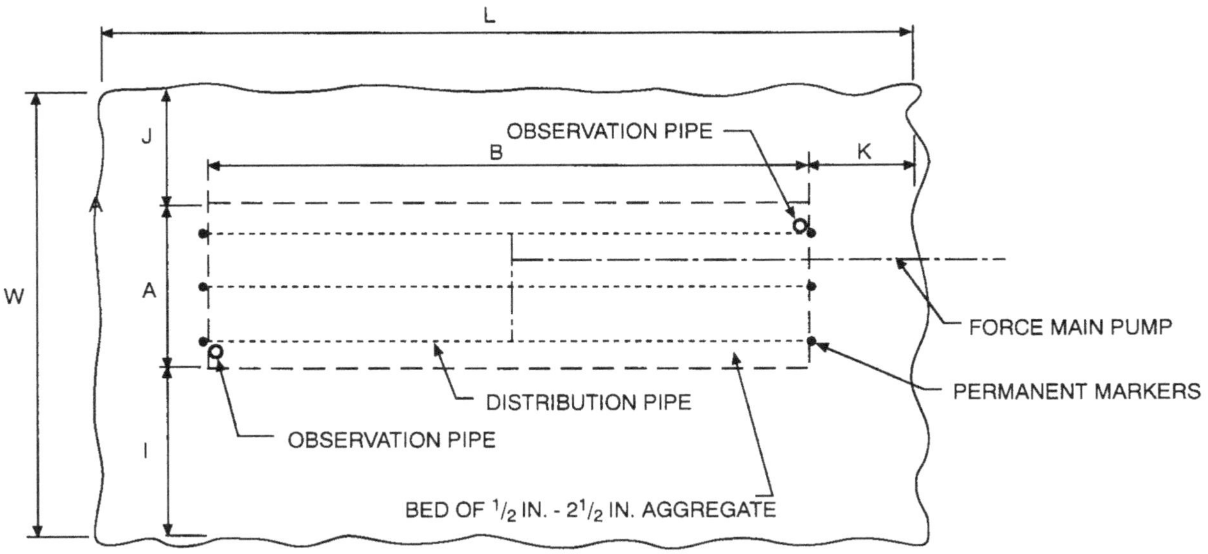

For SI: 1 inch = 25.4 mm.

FIGURE A-7 (SECTION 903.1)
PLAN VIEW OF MOUND USING A BED FOR THE ABSORPTION AREA

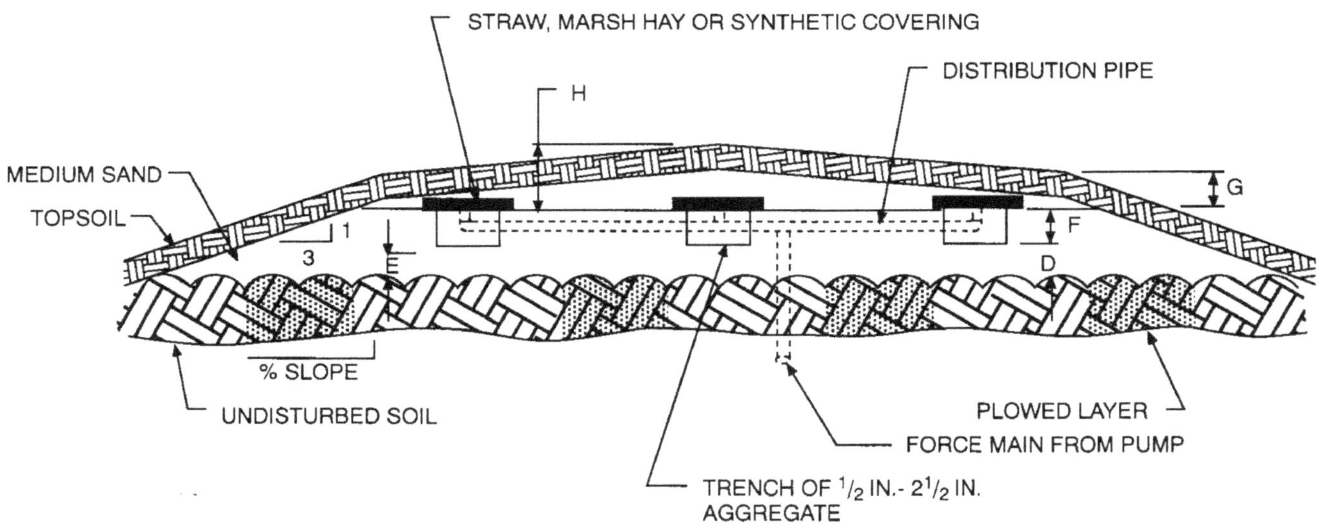

For SI: 1 inch = 25.4 mm.

FIGURE A-8 (SECTION 903.1)
CROSS SECTION OF A MOUND SYSTEM USING THREE TRENCHES FOR THE ABSORPTION AREA

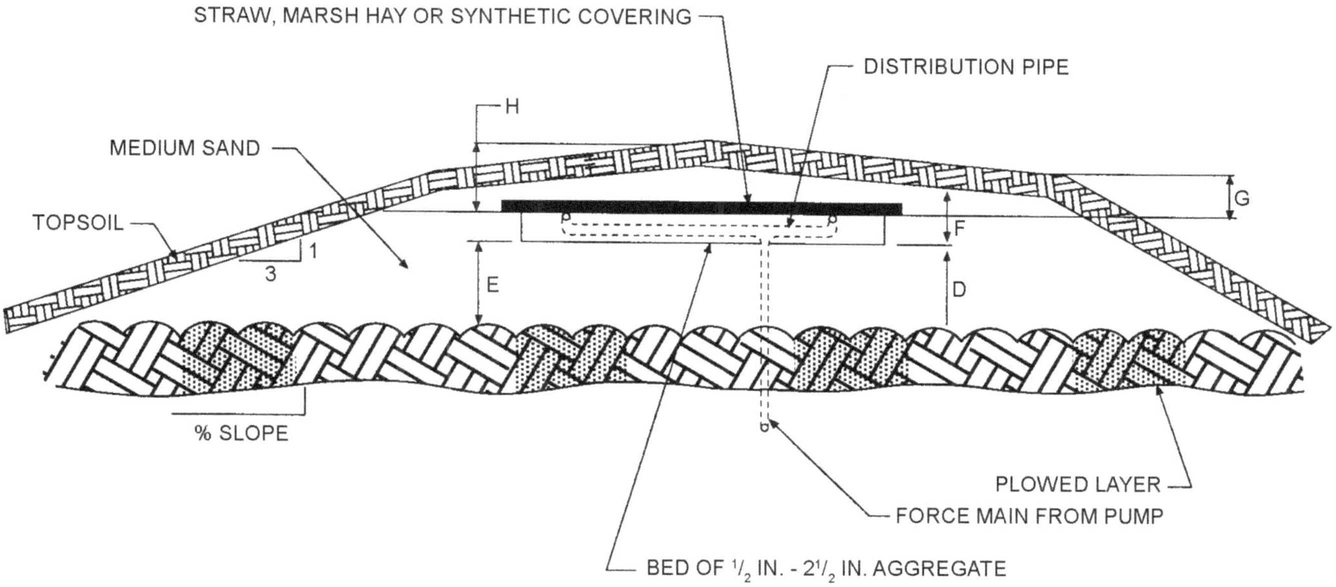

For SI: 1 inch = 25.4 mm.

FIGURE A-9 (SECTION 903.1)
CROSS SECTION OF A MOUND SYSTEM USING A BED FOR THE ABSORPTION AREA

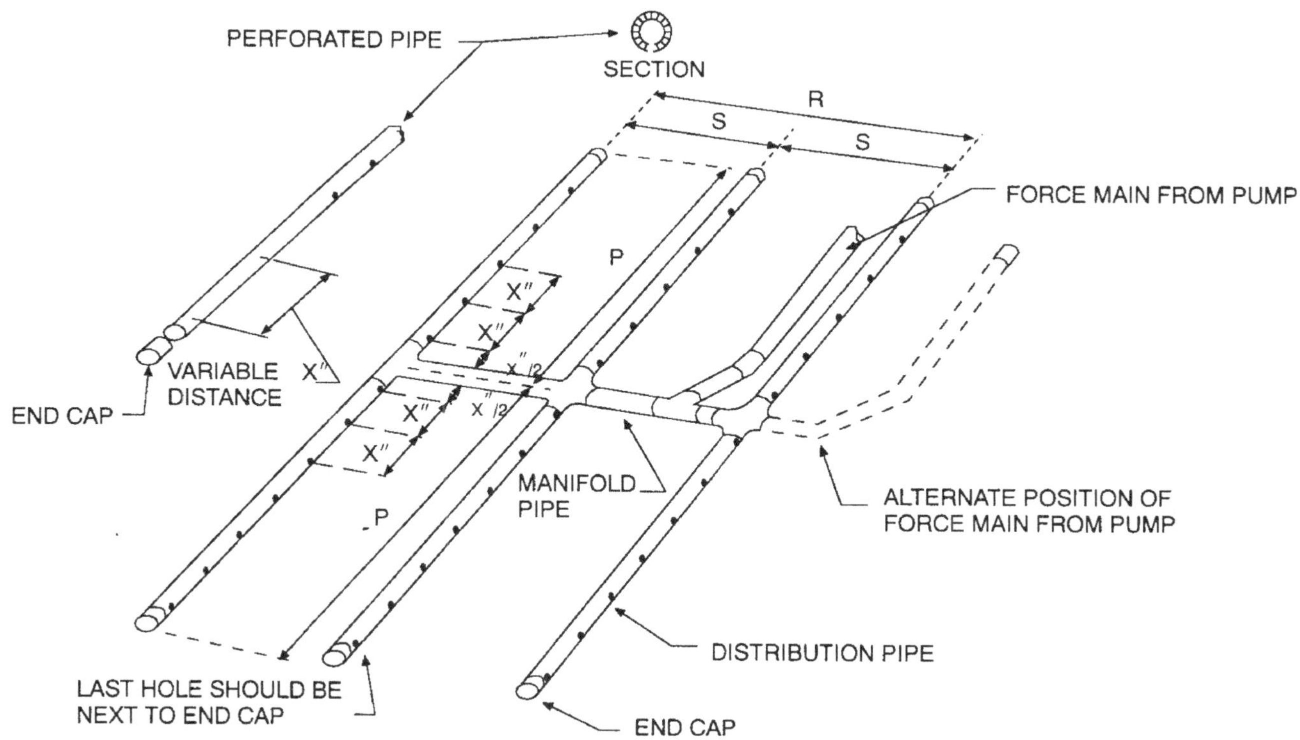

Note: Holes located on bottom are equally spaced.
For SI: 1 inch = 25.4 mm.

FIGURE A-10 (SECTION 903.1)
DISTRIBUTION PIPE LAYOUT

APPENDIX B

TABLES FOR PRESSURE DISTRIBUTION SYSTEMS

The provisions contained in this appendix are not mandatory unless specifically referenced in the adopting ordinance.

User note:

About this appendix: *Appendix B provides design nomographs and tables for the design of pressure distribution systems.*

TABLE B-1
REQUIRED DISTRIBUTION PIPE DIAMETERS FOR VARIOUS HOLE DIAMETERS, HOLE SPACINGS AND DISTRIBUTION PIPE LENGTHS (SCHEDULE 40 PLASTIC PIPE)

DISTRIBUTION PIPE LENGTH (feet)	DISTRIBUTION PIPE DIAMETER (inch)																													
	Hole diameter (inch) 1/4						Hole diameter (inch) 5/16						Hole diameter (inch) 3/8						Hole diameter (inch) 7/16						Hole diameter (inch) 1/2					
	Hole spacing (feet)						Hole spacing (feet)						Hole spacing (feet)						Hole spacing (feet)						Hole spacing (feet)					
	2	3	4	5	6	7	2	3	4	5	6	7	2	3	4	5	6	7	2	3	4	5	6	7	2	3	4	5	6	7
10	1	1	1	1	1	1	1	1	1	1	1	1	1	1	1	1	1	1	1	1	1	1	1	1	1 1/4	1	1	1	1	1
15	1	1	1	1	1	1	1	1	1	1	1	1	1 1/4	1	1	1	1	1	1 1/4	1 1/4	1	1	1	1	1 1/4	1 1/4	1 1/4	1	1	1
20	1	1	1	1	1	1	1 1/4	1	1	1	1	1	1 1/4	1 1/4	1	1	1	1	1 1/4	1 1/4	1 1/4	1	1	1	2	1 1/2	1 1/4	1 1/4	1 1/4	1
25	1 1/4	1	1	1	1	1	1 1/4	1 1/4	1	1	1	1	1 1/2	1 1/4	1 1/4	1 1/4	1	1	2	1 1/2	1 1/4	1 1/4	1 1/4	1 1/4	2	2	1 1/2	1 1/4	1 1/4	1 1/4
30	1 1/4	1 1/4	1	1	1	1	1 1/2	1 1/4	1 1/4	1 1/4	1	1	2	1 1/2	1 1/2	1 1/4	1 1/4	1 1/4	2	2	1 1/2	1 1/4	1 1/4	1 1/4	3	2	2	1 1/2	1 1/2	1 1/4
35	1 1/2	1 1/4	1 1/4	1	1	1	2	1 1/2	1 1/4	1 1/4	1 1/4	1	2	2	1 1/2	1 1/4	1 1/4	1 1/4	3	2	2 1/2	1 1/2	1 1/2	1 1/4	3	3	2	2	1 1/2	1 1/2
40	1 1/2	1 1/4	1 1/4	1 1/4	1	1	2	1 1/2	1 1/2	1 1/4	1 1/4	1 1/4	3	2	1 1/2	1 1/2	1 1/4	1 1/4	3	2	2	2	1 1/2	1 1/2	3	3	2	2	2	1 1/2
45	2	1 1/2	1 1/4	1 1/4	1	1	2	2	1 1/2	1 1/4	1 1/4	1 1/4	3	2	2	1 1/2	1 1/2	1 1/2	3	3	2	2	2	1 1/2	3	3	3	2	2	2
50	2	1 1/2	1 1/4	1 1/4	1 1/4	1 1/4	3	2	2	1 1/2	1 1/2	1 1/4	3	3	2	2	2	1 1/2	3	3	2	2	2	1 1/2	3	3	3	3	2	2

For SI: 1 inch = 25.4 mm, 1 foot = 304.8 mm.

TABLE B-2[a]
DISTRIBUTION PIPE DISCHARGE RATE

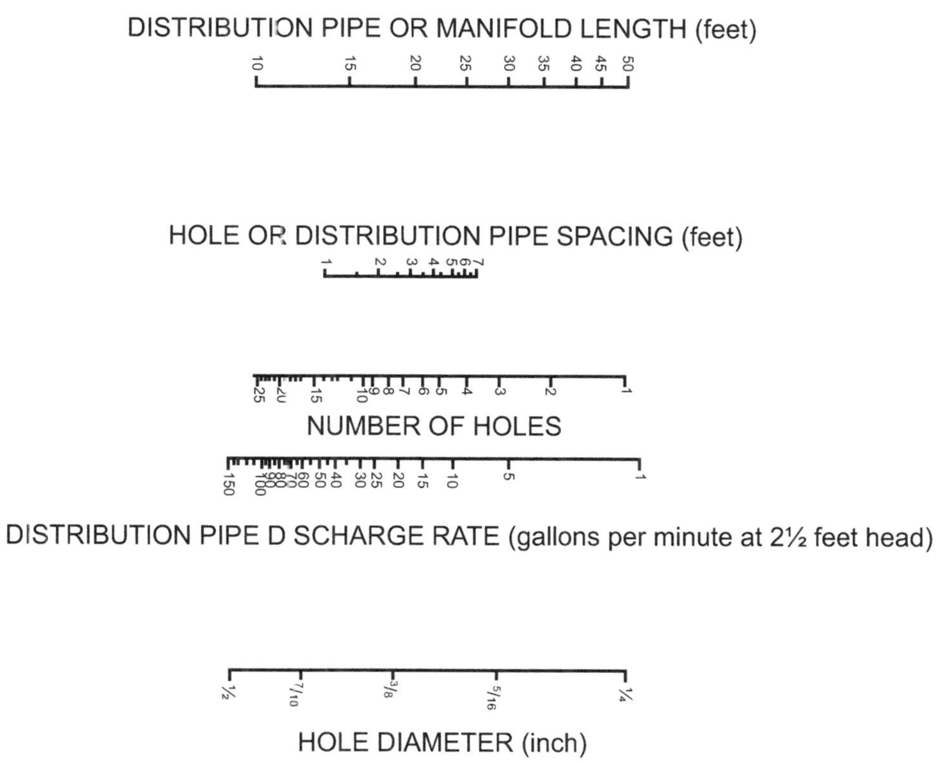

For SI: 1 inch = 25.4 mm, 1 foot = 304.8 mm, 1 gallon per minute = 3.785 L/m.

a. This table, a nomogram, determines the distribution pipe or manifold length, hole or distribution pipe spacing, number of holes, distribution discharge rate and hole diameter of pressure distribution systems by the placement of a straightedge between two known points.

TABLE B-3
RECOMMENDED MANIFOLD DIAMETERS FOR VARIOUS MANIFOLD LENGTHS, NUMBER OF DISTRIBUTION PIPES AND DISTRIBUTION PIPE DISCHARGE RATES (SCHEDULE 40 PLASTIC PIPE)

FLOW PER PIPE (gpm)	MANIFOLD LENGTH (feet): 5		10				15					20					25					30					FLOW PER PIPE (gpm)
Number of distribution pipes with central manifold	4	6	4	6	8	10	4	6	8	10	12	6	8	10	12	14	6	8	10	12	14	6	8	10	12	14	
	Manifold diameter (inch)																										
5	1	$1\frac{1}{4}$	$1\frac{1}{4}$	$1\frac{1}{4}$	$1\frac{1}{2}$	2	$1\frac{1}{4}$	$1\frac{1}{2}$	2	2	2	$1\frac{1}{4}$	$1\frac{1}{2}$	2	2	3	2	2	3	3	3	2	2	3	3	3	10
10	$1\frac{1}{4}$	$1\frac{1}{2}$	$1\frac{1}{2}$	2	2	3	2	2	3	3	3	2	3	3	3	3	3	3	3	3	3	3	3	3	4	4	20
15	$1\frac{1}{2}$	2	3	3	3	3	2	2	2	2	4	3	3	3	3	4	3	3	4	4	4	3	3	4	4	4	30
20	2	3	3	3	3	3	3	3	3	3	4	3	3	4	4	4	3	4	4	4	4	3	4	4	4	4	40
25	2	3	3	3	3	4	3	3	3	4	4	3	3	4	4	4	4	4	4	4	4	4	4	4	6	6	50
Number of distribution pipes with end manifold	2	3	2	3	4	5	2	3	4	5	6	3	4	5	6	7	3	4	5	6	7	3	4	5	6	7	

FLOW PER PIPE (gpm)	MANIFOLD LENGTH (feet): 35						40							45								50									FLOW PER PIPE (gpm)
Number of distribution pipes with central manifold	6	8	10	12	14	16	6	8	10	12	14	16	18	6	8	10	12	14	16	18	20	6	8	10	12	14	16	18	20	22	
	Manifold diameter (inch)																														
5	2	2	3	3	3	3	2	3	3	3	3	3	3	2	3	3	3	3	3	3	3	2	3	3	3	3	3	3	4	4	10
10	3	3	3	3	3	3	3	3	3	4	4	4	4	3	3	3	4	4	4	4	4	3	3	3	4	4	4	4	4	4	20
15	3	3	4	4	4	4	3	4	4	4	4	4	6	3	4	4	4	4	6	6	6	3	4	4	4	4	6	6	6	6	30
20	3	4	4	4	6	6	3	4	4	6	6	6	6	4	4	4	6	6	6	6	6	4	4	6	6	6	6	6	6	6	40
25	4	4	4	6	6	6	4	4	4	6	6	6	6	4	4	6	6	6	6	6	6	4	4	6	6	6	6	6	6	6	50
Number of distribution pipes with end manifold	3	4	5	6	7	8	3	4	5	6	7	8	9	3	4	5	6	7	8	9	10	3	4	5	6	7	8	9	10	11	

For SI: 1 inch = 25.4 mm, 1 foot = 304.8 mm, 1 gallon per minute = 3.785 L/m.

TABLE B-4[a]
PUMP DOSING RATE

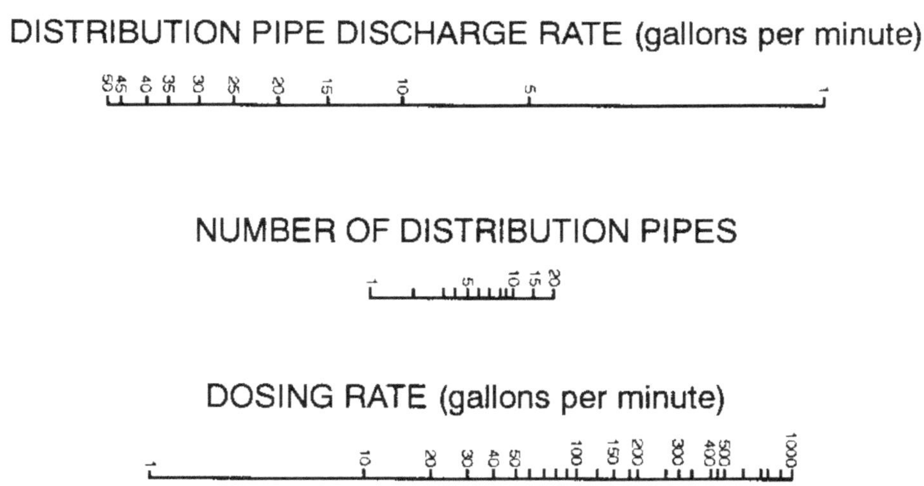

For SI: 1 gallon per minute = 3.785 L/m.

a. This table, a nomogram, determines the distribution pipe or manifold length, hole or distribution pipe spacing, number of holes, distribution discharge rate and hole diameter of pressure distribution systems by the placement of a straightedge between two known points.

TABLE B-5
FRICTION LOSS[a] IN SCHEDULE 40 PLASTIC PIPE (C = 150)

FLOW (gpm)	PIPE DIAMETER (inch)								
	1	$1^1/_4$	$1^1/_2$	2	3	4	6	8	10
1	0.07	—	—	—	—	—	—	—	—
2	0.28	0.07	—	—	—	—	—	—	—
3	0.60	0.16	0.07	—	—	—	—	—	—
4	1.01	0.25	0.12	—	—	—	—	—	—
5	1.52	0.39	0.18	—	—	—	—	—	—
6	2.14	0.55	0.25	0.07	—	—	—	—	—
7	2.89	0.79	0.36	0.10	—	—	—	—	—
8	3.63	0.97	0.46	0.14	—	—	—	—	—
9	4.57	1.21	0.58	0.17	—	—	—	—	—
10	5.50	1.46	0.70	0.21	—	—	—	—	—
11	—	1.77	0.84	0.25	—	—	—	—	—
12	—	2.09	1.01	0.30	—	—	—	—	—
13	—	2.42	1.17	0.35	—	—	—	—	—
14	—	2.74	1.33	0.39	—	—	—	—	—
15	—	3.06	1.45	0.44	0.07	—	—	—	—
16	—	3.49	1.65	0.50	0.08	—	—	—	—
17	—	3.93	1.86	0.56	0.09	—	—	—	—
18	—	4.37	2.07	0.62	0.10	—	—	—	—
19	—	4.81	2.28	0.68	0.11	—	—	—	—
20	—	5.23	2.46	0.74	0.12	—	—	—	—
25	—	—	3.75	1.10	0.16	—	—	—	—
30	—	—	5.22	1.54	0.23	—	—	—	—
35	—	—	—	2.05	0.30	0.07	—	—	—
40	—	—	—	2.62	0.39	0.09	—	—	—
45	—	—	—	3.27	0.48	0.12	—	—	—
50	—	—	—	3.98	0.58	0.16	—	—	—
60	—	—	—	—	0.81	0.21	—	—	—
70	—	—	—	—	1.08	0.28	—	—	—
80	—	—	—	—	1.38	0.37	—	—	—
90	—	—	—	—	1.73	0.46	—	—	—
100	—	—	—	—	2.09	0.55	0.07	—	—
125	—	—	—	—	—	0.85	0.12	—	—
150	—	—	—	—	—	1.17	0.16	—	—
175	—	—	—	—	—	1.56	0.21	—	—

(continued)

TABLE B-5—continued
FRICTION LOSS[a] IN SCHEDULE 40 PLASTIC PIPE (C = 150)

FLOW (gpm)	PIPE DIAMETER (inch)								
	1	1¼	1½	2	3	4	6	8	10
200	—	—	—	—	—	—	0.28	0.07	—
250						—	0.41	0.11	—
300						—	0.58	0.16	—
350						—	0.78	0.20	0.07
400						—	0.99	0.26	0.09
450						—	1.22	0.32	0.11
500	Velocities in this area become too great for the various flow rates and pipe diameter					—	—	0.38	0.14
600						—	—	0.54	0.18
700						—	—	0.72	0.24
800						—	—	—	0.32
900						—	—	—	0.38
1,000						—	—	—	0.46

For SI: 1 inch = 25.4 mm, 1 foot = 304.8 mm, 1 gallon per minute = 3.785 L/m.

a. Friction loss expressed in units of feet per 100 feet.

TABLE B-6[a]
MINIMUM DOSE VOLUME BASED ON PIPE SIZE, LENGTH AND NUMBER

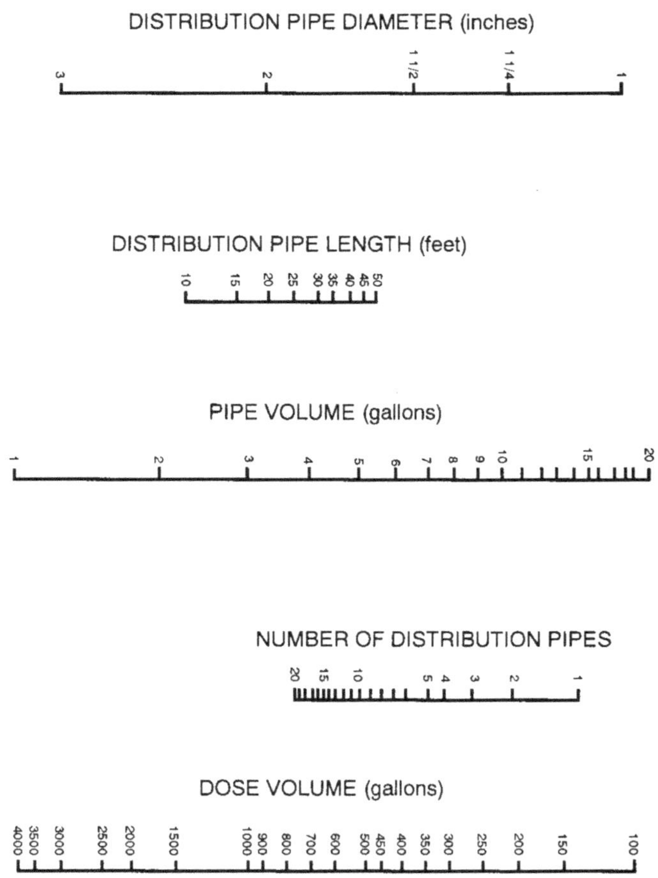

For SI: 1 inch = 25.4 mm, 1 foot = 304.8 mm, 1 gallon = 3.787 L.

a. This table, a nomogram, determines the distribution pipe or manifold length, hole or distribution pipe spacing, number of holes, distribution discharge rate and hole diameter of pressure distribution systems by the placement of a straightedge between two known points.

INDEX

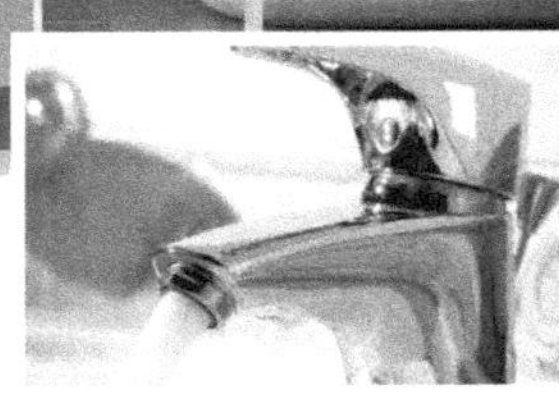